MICHIGAN BUSINESS
REPORTS *New Series* NUMBER 60

The Metal Finishing Industry
A Framework for Success

HAROLD E. ARNETT
DONALD N. SMITH

A publication of the
Division of Research
Graduate School of Business Administration
The University of Michigan
Ann Arbor, Michigan

Printed in the United States of America
ISBN–0–87712–182–6

CONTENTS

TABLES

FIGURES

METAL FINISHERS' FOUNDATION

The Metal Finishers' Foundation was founded on May 11, 1964 "to promote the educational and scientific advancement of metal finishing solely through the receipt of contributions and awards thereof to organizations qualifying under the Internal Revenue Code."

The Foundation's first project was an in-depth study of the metal finishing industry, published by the Bureau of Business Research of the Graduate School of Business Administration, The University of Michigan, in 1967, entitled "Business and Economic Evaluation of the Metal Finishing Industry." This was the first detailed study ever made of the industry.

Close to $700,000 has been expended by the Foundation since 1967 for business and scientific projects. One of the largest has been concerned with water pollution research and development programs, made possible by support from industry members and the Environmental Protection Agency. Another important undertaking was the development of a 16mm movie on "Metal Finishing - A Career Opportunity," an effort to attract new, dynamic employees into the industry.

Thus, the Foundation has made outstanding progress in serving the Metal Finishing Industry in the past and looks forward to continuing to do so in the future, to a large degree because of the outstanding leadership of its presidents.

1975-1977	*W. R. Grawford*
1974-1975	*E. N. Marlette*
1972-1974	*R. L. Springer*
1971-1972	*W. V. Turner*
1964-1971	*A. T. Leonard*

PREFACE

Everyone hopes for success. Many are willing to strive hard to attain it. It is not surprising, therefore, that the Metal Finishers' Foundation engaged us in September 1976 on a project to determine what the criteria are for a successful metal finishing firm.

In one sense this writing is a continuation and update of the 1967 study written for the same organization.* We made heavy use of that study in this monograph and thank the authors and the Bureau of Business Research for their cooperation in allowing us to make such liberal use of the information.

But we have attempted to do more. First, we have attempted to determine, to present, and to discuss those things necessary to operate a metal finishing firm successfully. Second, we have recommended approaches and procedures to be followed to attain that success. We have tried to be as specific in our coverage and recommendations as possible, while remaining mindful that the constraints of wide variability in the operations of different metal finishing firms and the attitudes and approaches of their managers make a high degree of specificity unattainable.

Success is an elusive term, meaning different things to different people. We feel our conclusions and recommendations, if followed, will aid managers in operating their firms more successfully and, perhaps of more importance, will better the performance of the industry in total. We recognize that all the recommendations will not be appropriate to each metal finishing firm, but we do hope they represent guidelines by which managers can gauge their performance. Beyond that, it is up to each manager to decide if he can and will adapt the suggestions to his own firm and then implement them.

* R. Lee Brummet and Harold E. Arnett, *Business and Economic Evaluation of the Metal Finishing Industry,* Michigan Business Reports, No. 52 (Ann Arbor: Bureau of Business Research, Graduate School of Business Administration, The University of Michigan, 1967).

We wish to thank the Foundation which sponsored this project and the representatives and members of the National Association of Metal Finishers who gave such enthusiastic support and assistance throughout, particularly those who responded to the questionnaire and those who welcomed us into their plants to study their operations and problems.

We also wish to express our thanks to John B. Arnett, candidate for the Master of Management degree from the Graduate School of Management at Northwestern University, for his assistance in collecting and analyzing information at various stages during the project.

Harold E. Arnett
Donald N. Smith

Ann Arbor, Michigan
February 1977

I

THE METAL FINISHING INDUSTRY
ITS NATURE, TRENDS, AND SIGNIFICANT PROBLEMS

What are the necessary ingredients for a successful metal finishing operation? This is the basic question answered by this research project. Searching for the answer has involved an analysis of the metal finishing industry — its sales, manpower, cost trends, and competitive problems. More specifically, we approached the problem simultaneously from opposite directions to see if we would arrive at the same conclusions.

First, we looked at the problems of the industry, feeling that if we could determine what the major problems were, we would then know what was necessary for a successful metal finishing operation; i.e., removing or at least reducing the major problems. From this direction, the investigation covered areas necessary to answer these questions:

1. What are the basic characteristics and trends of the industry, including the number of firms, their size, structure, location, employment, payrolls, and value of shipments?
2. What are the key competitive problems facing independent metal finishers?
3. How does the cost structure of independent firms influence their competitive position?
4. How do independent firms in the industry set prices, and to what extent do these practices influence competitive positions?
5. What are the primary economic and business justifications for continued existence and growth of independent metal finishing firms?

Second, we searched for the answers by looking at successful firms — analyzing their cost structure, pricing methods, organizational structure, and operations to determine what accounted for their success and how they avoided the pitfalls of the unsuccessful firms.

In order to obtain data to pursue these approaches toward a solution, we made an analysis of the general characteristics of the industry, gathering information from published sources, Association records, and selected field interviews. We interviewed the owner-managers of a number of independent finishing firms to secure the benefit of their experience and learn of their operating results. We visited some twenty-five firms in numerous locations, particularly in Michigan, Ohio, and California. The selection of these firms represented an attempt to cover the several types of finishing processes and finishes. This cross section of operations gave us the opportunity to observe the major types of activities and to determine the major problems of the industry as a whole. A questionnaire was sent to 750 firms. There were 150 questionnaire responses (20 percent). These responses from companies comprising perhaps 40 to 50 percent of the industry's output were very useful in pinpointing the major problems facing metal finishing firms, in indicating the weaknesses which need to be overcome if the industry is to maintain and improve its position, and in indicating what the successful firms are doing.

What Is Metal Finishing?

Finishing companies perform various services related to coating, polishing, buffing, or plating of components. Plating with various metals (gold, chrome, silver, zinc, brass, bronze, etc.), buffing, cleaning, descaling, anodizing, and painting are only a few of the many services offered. Independent finishers tend to restrict their services to one or two of these broad service categories.

For classification and analysis purposes, it has been convenient to organize information about the finishing industry into two broad categories. One covers those companies specializing in electroplating, plating, polishing, anodizing, and coloring; the other classification includes companies doing coating, engraving, and painting-related services.

Although companies within each of these two classifications share many of the same problems and opportunities, a more sensitive organization of industry data is required to measure growing specializations. As a result, three distinct groups now are clearly discernible:

1. Polishing and buffing
2. Coating
3. Plating

Plating is by far the largest group of the three, and electroplating comprises the major activity within that category.

Many factors determine the type of process which must be followed and the equipment and material which must be used to meet specific finishing requirements of a job. Chief among these are the part shape, size, and surface tolerance; the metal to be plated, buffed, or painted; the batch size and frequency of reruns; and whether the goal of the finish is for decoration, corrosion, wear protection, or conductivity. A complete list would virtually be endless when all materials and finishing requirements are consided.[1] In any case, there is an enormous array of individual services and combinations of services that are performed by the industry in total, and it represents a significant group of small businesses employing approximately 90,000 workers at the end of 1974.

Types of Metal Finishers

Metal finishing operations are performed by both captive and independently owned shops. A captive shop, as the term implies, primarily serves its parent company. Metal finishing is integrated into the manufacturing process of the parts or subassemblies being produced by the parent. Thus, metal finishing is only one part of a chain of activities performed by the company in order to convert raw materials into the final products and to sell them to the customer. An independent shop, on the other hand, is one whose primary function, and perhaps the only one, is to finish parts or products for others.

While precise data are not available, industry sources estimate that U.S. finishing business during the 1950s and 1960s was split between the captive and independent shops on approximately a 60–40 percent basis in favor of the captives. During the late 1950s and in the early 1960s, there was a strong trend for manufacturing companies to establish captive shops to meet their finishing needs. More recently, however, strict enforcement of water pollution laws has forced many manufacturers either to add expensive waste treatment facilities or to close their finishing operations. Increasingly, many are choosing the latter alternative and are resuming subcontracting of this work to independent shops. For independent shops in many industrialized centers, however, competition from captive shops continues to be a serious problem — one we briefly consider in Chapter VIII.

Characteristics of Independent Job Shops

Prior to the stricter enforcement of the water pollution laws, a simple finishing operation could be started in rented quarters for under \$10,000. As late as 1972, the average investment amounted to about \$60,000 for all established companies specializing in finishing operations, and about half that investment for polishing or buffing companies. Although the pollution abatement requirements and the need for labor

1. For those interested in a more complete description of finishing processes, a number of sources of information are available, such as the American Society for Metals, *Metals Handbook,* 8th ed., vol. 2 (1964), particularly the chapters on "Heat Treating, Cleaning and Finishing."

saving automated equipment to combat inflating labor costs are raising investment requirements, the relative ease of entry into the industry has encouraged many technicians to exploit their finishing know-how and start their own businesses. A natural consequence is that many of these new managers are more oriented toward technical considerations than toward business management.

Most independent shops began and have continued as family operations. If they are not family proprietorships, they are either partnerships or closely held cooperations. The owners tend to be independent, hard working, strongly individualistic, and highly motivated by a belief in free enterprise.

According to U.S. Department of Commerce data, there are about 4,700 independent finishing companies; 3,200 concentrate on polishing and plating, while 1,500 specialize in coating. In the plating and polishing segment, 92 percent of the firms have less than 50 employees, compared to 88 percent of the coating companies.

Market penetration data disclose different relationships from those suggested in the distribution of the companies classified by size. While coating companies of less than 50 employees constitute 88 percent of the industry, they account for only 36 percent of the total output. Similarly, the 92 percent of the plating and polishing companies employing 50 workers or less account for just 61 percent of the output, as the following figures reflect.

Fig. 1. Share of Market by Company Size Plating and Polishing Companies

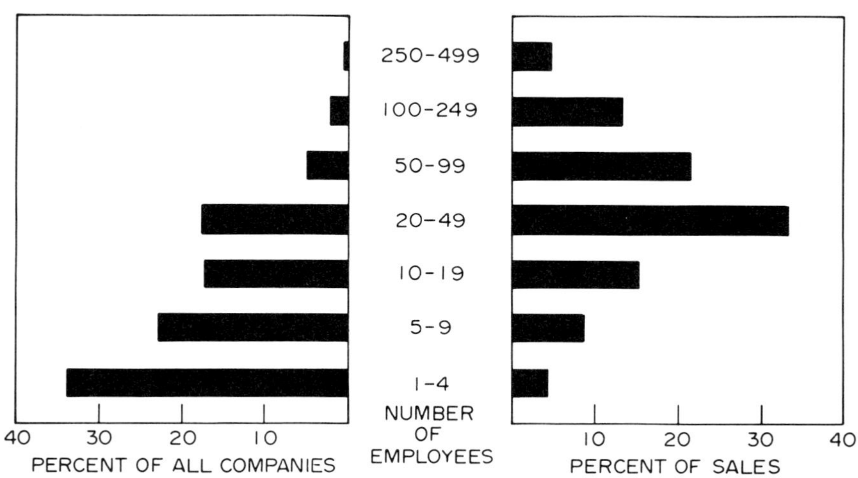

Fig. 2. Share of Market by Company Size
Coating Companies

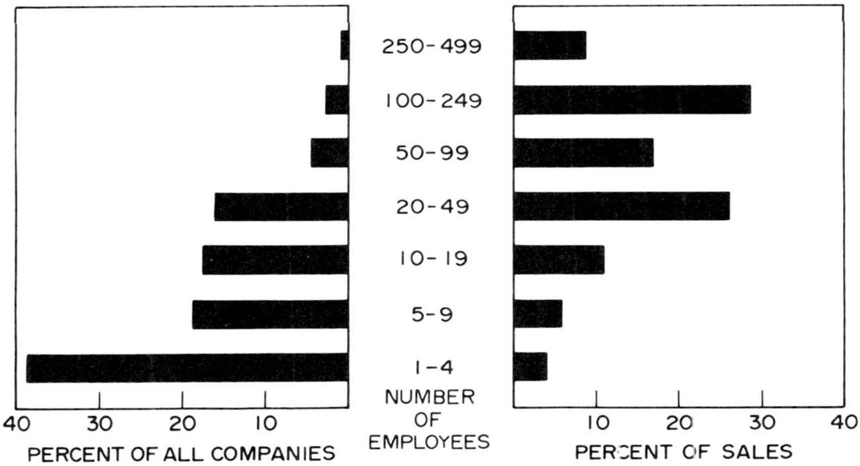

There has been only a slight expansion in the size of the typical shop during the past two decades. Although sales in the industry grew from $400 million in 1954 to about $2 billion in 1973 (the latest year that U.S. census data are available), the proportion of companies having 20 or more employees expanded only slightly — from 20 to 24 percent. Forty percent still have fewer than five employees.

The barriers constraining the size of firms in the industry have produced a group of small companies fiercely competing with each other and with captive shops. This has resulted in a spirited level of competition among many bidders for contracts, relatively low profits, and undercapitalized operations. The problems evolving from the size structure of the typical shop are serious. Note this statement of one independent shop owner:

> We are an industry of small businesses sandwiched between a supplier industry, on the one hand, characterized by giants, and customers, usually of equal size, on the other . . . . In many years our materials suppliers have slapped us with semiannual cost increases of 15 to 20 percent each . . . . When we try to pass along part of these cost increases to our customers in the form of 2 to 3 percent price increases per year, they refuse to go along.

Still, the independent operator's ability to resolve such challenges successfully creates the opportunity and framework for a successful independent finishing industry. The history of the industry shows that its development has been the result of highly motivated entrepreneurs seizing on the production problems of a nearby manufacturer and skill-

Fig. 3. Growth of the Finishing Industry

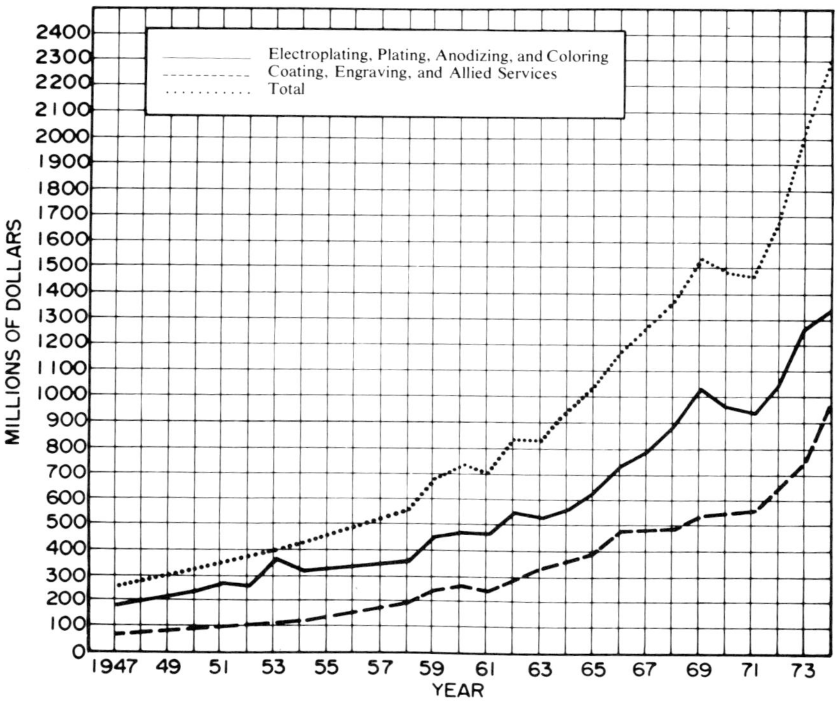

fully engineering a more economical or superior solution to his finishing problem. The finishing firm manager who is able to resolve another company's finishing needs at consistent profits will be able to generate the necessary working capital for growth and for the acquisition of modern automated equipment.

The quest for monetary rewards and the desire for "independency" inherent in working for oneself entice many to begin on a "shoestring" by establishing the minimal activity. This typically is a single line or a couple of buffing machines. The idea to start a business often arises from perceiving a finishing need in a company that may perhaps employ an acquaintance. When initial attempts are successful, or if there is open capacity, new customers are sought. This usually leads to expansion. Expansion characteristically occurs through the addition of more manual or automated plating, coating, or painting lines, or additional or more sophisticated polishing machines. Seldom is expansion achieved by adding a second or third plant; multi-plant independent finishing companies are rare.

Location of Plating Industry

Because independent finishers need to be near their customers and most of their customers are manufacturers of durable goods, the dispersion and clustering of the industry roughly parallels the locational patterns of U.S. durable goods producers. The leading metal working states, specifically California and the states in the North East and North Central regions, understandably have most of the finishing companies—approximately 82 percent of the industry total. Specific data on the six leading metal finishing states are shown in Table 1.

Companies in the heavily industrialized states of Illinois, Michigan, and Ohio have attained the largest average size in terms of employees. It is significant that productivity, as measured by the value added per worker, is greater for these companies; that is, productivity appears to increase as the size of the firm increases. This suggests that low profits and fierce competition in the industry have prevented companies from reaching the size necessary for maximum production economies of scale. Consequently, productivity, as generally measured, has been retarded. Certainly this also is a major problem of the industry.

Competition in the metal finishing industry is intense, not only because of the relative ease of entry into the industry, which accounts for the large number of firms, but also because the shops tend to cluster in given areas. It is not unusual to find as many as 200 to 300 finishing plants in large metropolitan areas such as New York, Detroit, Chicago, and Los Angeles. Ninety-one percent of questionnaire respondents indicated there are many competitors in their area.

Market Characteristics

Company concentrations are a natural outgrowth of the service requirements of the finishing customers. Independent shops locate close to the plants they serve in order to minimize delivery time and cost. More than three-fourths of a typical finishing shop's business comes from an area within a radius of 50 to 75 miles.

Parts to be finished are almost always moved from the customer's plant to the finisher and then transported back. Often the ability of a customer to send a part out for finishing and get it back the same day decides who, among several economically priced bidders, gets that company's business.

Metal finishing is one of the last functions performed before product assembly and shipping. In fact, customers whose production schedules are running behind often expect the finishing operation to be done quickly enough to make up at least part of the lost time. Customers sometimes demand unreasonable delivery schedules from the finisher because their own customers are pressing for quick delivery. Such

TABLE 1

LEADING METAL FINISHING STATES

State	No. of Firms	Employees	Average Size of Firm (Employees)	Value Added ($ Millions)	Average Value Added Per Firm ($ Thousands)	Value Added Per Worker
California	742	11,600	15.6	148.1	200	$12,767
Illinois	445	9,500	21.4	141.7	318	14,915
Michigan	455	8,700	19.2	122.8	269	14,114
Ohio	389	7,100	18.3	105.5	272	14,859
New England	510	7,900	15.4	98.0	192	12,405
New York	494	7,200	14.6	86.0	174	13,502
Totals	3,035	52,000	17.4	702.1	$237.5	

Source: *U.S. Census of Manufacturers*, 1972.

situations create scheduling conflicts, and, if they are not resolved diplomatically, customers may go to other finishers. In more extreme situations, they may set up their own captive shops.

Because the industry's major customers commonly manufacture products that are classified as consumer or producer durables, individual finishers can experience substantial swings in the demand for their services. For example, when the national economy turns down, durable goods producers, such as those in appliance, furniture, equipment and machinery, and automotive industries, frequently experience declines in sales greater than those for the economy as a whole.

Even though the major share of their output goes to these relatively volatile industries, the total sales pattern of the finishing industry reflects less severe fluctuations than those in many other industries (particularly capital goods producers) which serve these same markets. This suggests that a finishing company can attain reasonably stable sales through customer diversification.

TABLE 2

HISTORICAL SALES PATTERNS OF METAL FINISHING INDUSTRY

Some examples of major customers and their products

Aircraft:	Gears, pins, compressor blades, turbine buckets, piston rings, engine parts
Automotive:	Wheel covers, instrument panels, bumpers, hardware and trim, bearings, handles, engine parts
Housing Hardware:	Plumbing fixtures, home and building lockplates and hinges, door knobs, cabinet trim
Household Appliances:	Trim items, trays, shelves, handles, frames, knobs
Furniture:	Trim, handles, tubular components, hinges
Motorcycles and Bicycles:	Handlebars, levers, wheels, sprockets
Tools and Dies:	Gages, instruments, machine slides, molds
Electronic Equipment:	Components for telephones, radios and televisions, computers, and instruments and connectors
Steel Mill Products:	Strip, wire, sheet, conduit
Costume Jewelry:	Wrist watch bands, earrings, pins, necklaces, rings

(It should be pointed out that this list barely scratches the surface in terms of the industries that use plating services.)

Major Problems

In response to the questions "What do you consider to be the major problems of your firm?" and "What do you consider to be the major problems of the metal finishing industry in general?" members of the Association answered as shown in Table 3.

A number of interesting observations regarding these data are possible, particularly compared with the findings in the 1967 study.[2] As to the firm, the 1967 respondents listed increasing costs as the fifth most important problem facing them. By 1976, increasing costs had become

TABLE 3

PROBLEMS CITED BY SURVEY COMPANIES

Major Problems	Total Number of Times Mentioned	Number of Times Listed as First Priority
Of the Firm:		
1. Increasing costs of production (materials, labor, etc.) and inability to pass these on to the customer	61	28
2. Lack of experienced and competent help (including drop in employee loyalty to the firm, inability to attract young, dynamic people to the firm, etc.)	61	21
3. Cut-throat competition—Competitors pricing below costs because of lack of knowledge of costs (including captive shops)	57	15
4. Lack of customer understanding of the capabilities, limitations, and problems of the industry (unreasonable customer demands regarding time allowed for job, quality of work performed, etc.)	57	4
5. Lack of capital to expand and modernize	54	4
6. Quality control (inability on the part of your firm to prevent inefficient utilization of cost factors, i.e., using more materials than reasonable, more labor than reasonable, etc.)	54	1
7. Inability to keep up with changes in the market (i.e., type of plating needed, quality of plating needed, etc.)	49	0
8. Pressures from organized labor	48	0

2. *Ibid.*, p. 4.

TABLE 3 *(Continued)*

PROBLEMS CITED BY SURVEY COMPANIES

Major Problems	Total Number of Times Mentioned	Number of Times Listed as First Priority
Of the Industry in General:		
1. Increasing costs of production (materials, labor, etc.) and inability to pass these on to the consumer	59	21
2. Cut-throat competition—Competitors pricing below costs because of lack of knowledge of costs (including captive shops)	58	21
3. Lack of experienced and competent help (including drop in employee loyalty to the industry, inability to attract young, dynamic people into the industry, etc.)	56	21
4. Lack of customer understanding of the capabilities, limitations, and problems of the industry (unreasonable customer demands regarding time allowed for job, quality of work performed, etc.)	54	5
5. Lack of capital to expand and modernize	53	4
6. Quality control (inability on the part of the metal finishing industry in general to prevent inefficient utilization of cost factors, i.e., using more materials than reasonable, more labor than reasonable, etc.)	52	4
7. Inability to keep up with changes in the market (i.e., type of plating needed, quality of plating needed, etc.)	50	2
8. Pressures from organized labor	50	1

the most important problem. This is explained largely by the severe inflation over the intervening years. As prices increase, it becomes increasingly important to use the cost factors efficiently and effectively if the firm is to be operated profitably. This accounts for quality control rising from a problem hardly mentioned in 1967 to one of significance in 1976. In reply to the question "Do you have problems with quality control?" 74 percent of questionnaire respondents answered affirmatively.

Lack of experienced help and cut-throat competition remained high on the list of problems in 1976, as they had been in 1967, though surpassed in this study by increasing costs. Pressures from organized labor remained high on the list of problems for certain firms, but since

1967 it has dropped significantly as a problem for a large number of firms. In fact, by 1976 it was mentioned by about 50 percent of the respondents as being last in a list of eight major problems.

We were somewhat surprised to find that lack of capital to expand and modernize dropped from the third most important problem for individual firms in 1967 to the fifth or sixth major problem by 1976. Apparently the pressure for pollution abatement equipment has not as yet had its full impact. We quizzed the respondents to the questionnaire as to the extent pollution abatement requirements are affecting their firms. Answers varied from "no effect at all" to "driving me out of business." Most answers fell in the range of moderate to strong effect. Obviously, these variances in answers reflect differences in enforcement between states. If the Environmental Protection Agency does invoke and enforce stringent effluent standards, we feel that lack of capital will become the most significant problem for individual firms and the industry.

For the industry in general, answers from the respondents almost exactly paralleled the answers given for their individual firms, with one or two relatively minor exceptions. Increasing costs of production, cut-throat competition, and lack of experienced and competent help were considered nearly equal in importance for the industry in general, while ranking in a somewhat different order for individual firms. Quality control was somewhat less important as a problem for individual firms than lack of capital to expand and modernize, while the opposite was true for the industry in general.

Thus, in 1976, respondents generally perceived the industry to have the same problems as their individual firms, with approximately the same relative importance. This was not the case in the 1967 study, referred to earlier. In 1967, respondents felt the overwhelming problem for the metal finishing industry in general was unfair competition, while the most important problem for the individual firms was lack of experienced and competent help. Nevertheless, most of the problems still persist even though they might have changed in relative importance over the intervening ten years. Since respondents perceive the problems of the industry in general and their specific firms in particular as being nearly the same, it should be somewhat easier for the Association and its members to implement programs to ease the problems.

Field interviews and discussions pointed out other significant problems, particularly with regard to the industry in general, which were not specifically covered on the questionnaire.

Many managers, for example, were not aware of the extent to which productivity has decreased in the industry during the last two decades. Although weak in certain aspects as a measurement tool, changes in output per man-hour provides an often used indication of productivity

Fig. 4. Productivity Gains of Certain Selected Industries Compared with the Finishing Industry

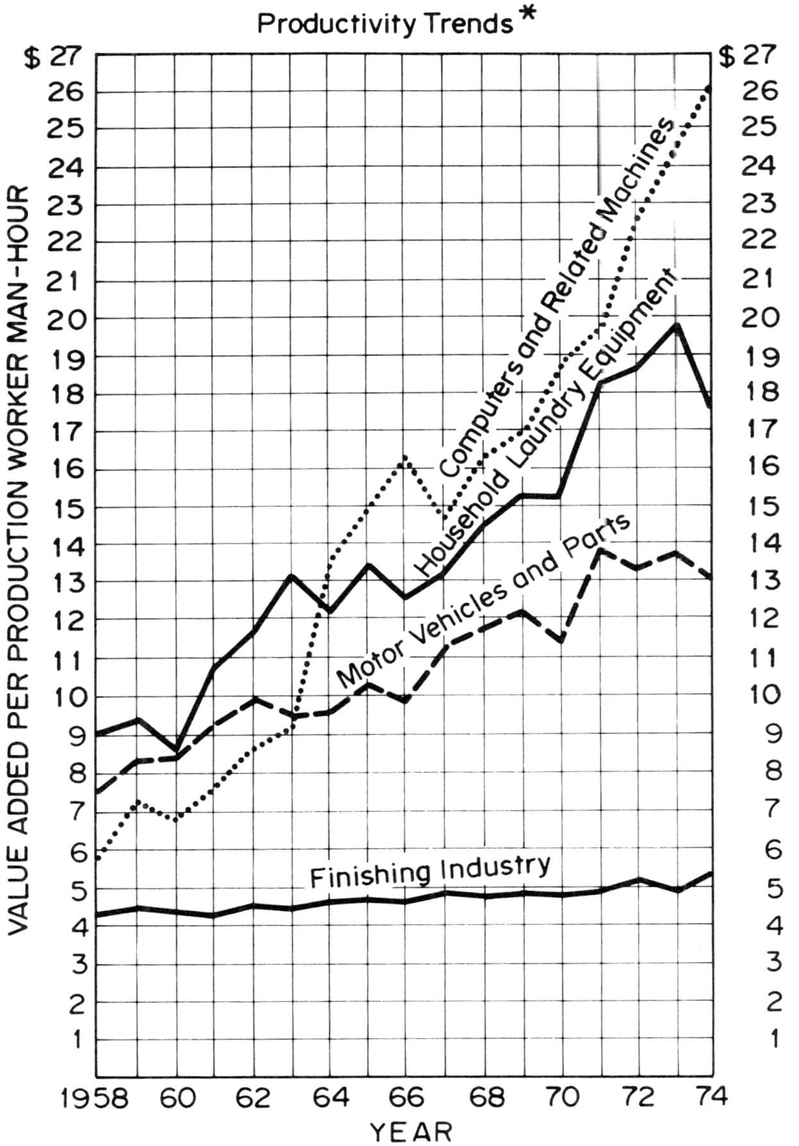

* These data are adjusted for inflation by using the implicit deflators (an index used to convert current dollar value to constant dollar value - base year 1958 = 100) from the U.S. Department of Commerce, Bureau of Economic Analysis workfiles

trends. The preceding figure dramatically discloses the decrease in productivity in the finishing industry relative to certain industries it serves.

In general, there are two distinct finishing industry patterns, and they have countertendencies. Some companies, dominated by manual operations, have actually experienced a decline in worker productivity over the past 10 to 15 years. By contrast, companies adapting or expanding automated equipment have enjoyed dramatic gains.

Where productivity has stagnated, it results from several factors. A simplistic analysis discloses an inadequacy in the refinements of manual finishing methods to overcome losses in output due to ongoing liberalization of work rules and schedules. Examples of the latter include increased break times during work periods, increased vacation and sick time, and other forms of paid time-off (all examples of concessions won by workers throughout the industry over the past several decades). There are also the dampening effects from generally increasing labor turnover and the greater utilization of marginally qualified workers.

One reason why switching to automatic production methods has been slow is lack of capital. Relatively low profit levels have made it impractical to attract capital through the sale of stock to the public. As a result, most independent firms are restricted to operating as family-held proprietorships. When outside funds are attracted for growth or equipment modernization programs, they often are in the form of bank loans secured by the assets of the company plus the personal assets of the company's owner. It is the nature of the metal finishing industry more than the conservatism of bankers that accounts for such collateral requirements. Should default of a loan occur, the lending bank faces limited prospects in liquidating the collateral. The current assets are needed to take care of ongoing operations, and the liquidation value of fixed assets frequently is only a fraction of its original price.

Besides encountering rapid deterioration from the corrosive chemicals, much of the equipment is affixed to the facility and costly to move. One company with equipment having a book value of $80,000, spent $35,000 to dismantle, move one block, and install it in a new facility. As a result, used finishing equipment, even when in sound condition, seldom commands more than 15 to 30 percent of its original cost.

The difficulty that metal finishing firms experience in amassing adequate capital has made it difficult to move from an industry characterized by labor intensive operations to one enjoying the enhanced productivity and quality benefits inherent in technologically advanced equipment.

Strapped by capital shortages, these small businesses also find it difficult to develop the capital to fund research and development. Almost all of the small firms and many of the larger ones are without research activities. In periods of rapidly changing technology such as this nation has experienced, effective research and development is indispensable for the successful development of new processes, and for the enhancement of productivity and quality. It is noteworthy that those firms which do engage in research and development activities are generally the most profitable. Besides being able to handle a variety of demands by their customers, they have better quality and output per worker.

In the absence of technological advances generated within the finishing industry, the suppliers of the industry have generally created the radical changes. Fortunately for the small job shops, a supplier with a new concept is quick to educate the industry about new equipment, materials, or techniques. However, the process of information dissemination of these advances has resulted in a temporarily more favorable position for the captive and large independent shops because of their buying power. Note these statements by two managers.

> "Because of the receptivity to new ideas by the larger finishing firms which have the available capital to implement the new technologies, the suppliers tend to throw the weight of their technical support and research organizations behind the largest plating operations. The sales/technical forces seem to focus on the large captive operations in particular. Production executives in these operations expect good technical service as part of purchasing a proprietary process."
>
> "The major suppliers are also very large companies which find fewer institutional barriers in cross market transactions with the larger firms that operate their own captive facilities. Future developments in methods can be expected to come from these suppliers. The collective pressure of small job shops for special processes is insignificant compared to that of the large captive shops."

The Foundation is encouraged to continue to provide, and even expand, valuable leadership on research problems.

In cases where capital could be obtained, shops compensated for the relaxations of work schedules and rules by replacing labor with automated equipment. Companies that raised the capital enjoyed dramatic productivity gains, sometimes expanding worker output three to five times over a period of a few years.

Other factors besides capital shortages have restricted wide scale adoption of automation. Chief among these are the sporadic and frequently unpredictable demands of finishing customers which periodically can leave a shop with idle capacity. The harsh economic reality is that it is less devastating to furlough or reassign workers in a manual operation than it is to absorb the high capital charges flowing from costly machinery and equipment sitting idle. Expensive automated lines that were frequently idled have bankrupted many job shops, abruptly throwing all the employees out of work. Other factors such as small batch sizes or short production runs, which characterize the finishing markets of many job shops, also militate against the efficient utilization of automated finishing equipment.

Because of these and many other less obvious factors, the finishing industry has generally found it difficult to enhance productivity. As can be seen in Figure 4, previously presented, Value Added by Manufacture Per Production Worker (which is essentially sales minus the cost of purchased items, in terms of constant dollars) shows that "real" value added increased just 71 cents per hour from $4.29 in 1958 to only $5.00 by 1973. This is slightly less than a one percent annual improvement.[3]

The preferential use of another index might produce minor arithmetic differences in the "real" annual percentage change of output, but the indications that a lagging productivity condition exists are clear. Productivity gains in the finishing industry not only have been a fraction of those achieved throughout the U.S. economy as a whole, but also only a fraction of the gains made by their principal customers, such as autos, appliances, and electronics manufacturers (see Figure 4).

The future prospects are also less than optimistic. Much of what limited capital the job shops do generate will be required to buy government mandated pollution abatement devices, so that little will be left for more productive finishing equipment.

New workers must also be added to maintain and operate the pollution controls. And since these workers will not be directly involved in finishing production, their addition to a plant's work force will have the effect of reducing the average output per worker for the plant. Then too, many plants attempt to meet the more stringent pollution laws by extending or modifying plating or rinse solutions and processes, or by switching to less polluting, but often less efficient, chemicals. For example, one company, in order to keep from being closed by the Environmental Protection Agency, sharply curtailed the cyanide in its process at the expense of doubling the total plating cycle time. Another company extended by fourfold the dwell time that a barrel of parts is held over the tank after removal from plating to allow more of the plating solution to drop back into the plating tank, thereby reducing pollution problems in subsequent solutions. A consequence of such adjustments sometimes is less efficient use of the line, and consequently lower productivity per attending worker.

Uncertainty about EPA regulations is also forcing many companies, both independent and captive shops, that do have the funds for more productive equipment to delay acquisition until they can be certain that their new process will meet future pollution standards. The government, through the Small Business Administration, or in other ways,

3. The deflator factor used in this analysis to adjust the Output dollars for the effects of inflation is one specifically computed by U.S. Department of Commerce, Bureau of Economic Analysis, for the Finishing Industry (SIC 3471). While this deflator is thought to be reasonably indicative of inflation in this industry, other analysts may prefer another deflator, such as the Wholesale Price Index or the Consumer Price Index.

is either going to have to make funds available for borrowing or stand ready to guarantee bank loans to buy pollution abatement equipment; otherwise, large numbers of finishing firms are likely to go out of business.

A final problem, and the most severe one in many ways, is that the managers of many firms, particularly the small ones, do not use accounting information effectively in operating their firms. In some cases they simply do not understand the informational content of the data presented to them. In other cases, different and better information is needed. In some instances, managers believe that they can operate their firms "by the seat of their pants"—i.e., that accounting information is not needed to operate successfully. In any case, greater use of better accounting information is needed if managers hope to enhance profits. In view of its considerable impact on profitability, the total area of cost analysis, general accounting, work-order control, budgeting, forecasting, pricing procedures and practices, needs to be developed and refined by many firms. Of prime importance is the efficient and effective feedback of the results produced by the accounting system so they can be used in the decision-making process.

All the problems mentioned will be considered again and in greater detail in later chapters, except quality control. Without doubt, the metal finishing industry in general, and specific firms in particular, need to exert considerable effort to establish standards for material, labor, and overhead utilization if it is to operate successfully. Such effort, in fact, is long overdue. However, because of the breadth and complexities of such an undertaking, the Metal Finishers' Foundation felt that quality control should be excluded from this study, perhaps to become the topic of a future research project.

II

A FRAMEWORK FOR SUCCESS AMONG SURVEY COMPANIES

Several on-site examinations of finishing plants and many interviews were conducted to determine what constitutes a framework for success.[1] Superior performance, we found, is not a question of performing the obvious functions expertly, such as high quality finishing completed on a timely basis. These functions must be performed satisfactorily, without question, if the firm is to continue in existence. And we found no survey firms which had operated for more than a few years that weren't performing such basic functions satisfactorily. What raises the successful firms above those just managing to survive, however, is sufficient management talent to meet successfully the innumerable problems that arise in marketing, engineering, production, finance, pricing and cost control.

In other words, in order to survive, a firm has to effectively meet the challenge of quality finishing on a timely basis. But a firm having significant idle capacity, or using too much material or labor, pricing incorrectly, or supervising workers poorly still might perform basic finishing functions effectively. Superior performance is attained by those firms that perform those functions *efficiently* — using no more materials and labor than necessary, aggressively obtaining sufficient jobs to keep operating activity at a high level, and pricing work correctly.

1. Observations regarding the criteria constituting a framework for success are drawn from data gleaned during the survey phase. Thus, these assessments relate directly to the members of the survey group. Nevertheless, many observations undoubtedly fit much of the industry. Still, universal inferences from any survey data would be valid only after scientific selection of a random sample from the entire industry, followed by acceptable survey research techniques.

Team Management

Meeting the many management challenges requires more executive skill and more human energy than usually possessed by one person. The single most discernable factor in companies that achieve above average success is the presence of more than one executive specializing in those functions he or she is most capable to perform. Successful companies often are directed by a father-son management team or, in a few instances, by a father-son-grandson trio. In one highly successful company, management participants include the founding husband and wife plus their two sons and a daughter.

Many factors account for the success of family teams, but certainly one important explanation is that owner-managers demonstrate a commitment to a company's success that far exceeds the level normally obtainable from employee-managers, regardless of the financial incentives involved. Pride in ownership seems to instill a willingness to work long periods in order to achieve company success, even when only modest salaries result and/or the return earned on investment is lower than exists in many other industrial sectors.

For companies managed by only one executive, a number of management tasks simply cannot be handled to the extent required for success, even in the smaller firms. Time is usually spent on "middle management" funtions — production scheduling, production control, inspection and quality control, monitoring of metal and chemical contents, design of racks, optimal part loading arrangements, meeting unrealistic customer schedule demands—leaving little if any time to perform "top management" functions, such as finance, pricing, marketing, and process research and development. When such functions are slighted, poor overall performance usually results.

Sixty-three percent of questionnaire respondents indicated that they performed all of the major activities of their firms by themselves. When all these demands are thrust upon the shoulders of a single executive, it is no surprise that satisfactory performance is difficult, perhaps impossible. By comparison, a management team, built upon family bonds or upon the association of two or three compatible but unrelated founding partners, each specializing in those activities he or she does best, is much more likely to meet the diverse challenges and operate successfully.

Total team management also encompasses the effective appointment and supervision of foremen. Two shortcomings relative to new foremen appointments were observed in the less successful companies. A mistake that many companies make when promoting a finisher from the ranks to foreman is to fail to provide adequate communications to other employees regarding the foreman's authority and responsibility. In making the promotion, owner-managers seek freedom from day-to-day production line problems so they can, for example, expand sales efforts by personally calling on new customers. Yet more than one manager indicated that upon returning from appointments away from the plant, they sometimes found worker unrest, occasionally severe. Unionization frequently follows.

Chances of such trouble can be lessened by providing clearly written directions to employees about the appointment. When designating a new foreman, successful finishing companies use written communication (1) to specify what they expect of the new foreman, (2) to delimit his areas of responsibility and authority, and (3) to emphasize that the new foreman speaks for the owner-manager, and that his rational actions will be backed by the owner. Subsequent employee-owner discussions of worker reactions also help to resolve negative reactions.

More is needed to make a new foreman a good manager. The industry commonly promotes the most proficient and industrious worker to the foreman position. Just because a finisher is a good technician, however, does not mean that he will be a good motivator of people. Too little consideration has been given to the potential foreman's ability to relate to, communicate with, and most important of all, to motivate the other workers. Rarely have workers been educated in the fundamentals of human relations and worker motivation. The performance of foremen throughout the entire industry could be enhanced greatly by formalized professional development and continuing education programs in all phases of personnel and financial management. In the less successful firms, in particular, this need was obvious.

A sharp contrast exists between the quality of the foremen in the successful firms compared with the struggling companies. First, many of the less successful companies do not have a foreman — production supervision may be the full responsibility of the owner-manager. Among the companies having foremen, the more profitable finishers successfully delegate most of the day-to-day routine production responsibilities to a foreman. Obviously, these supervisors earn this broadened authority by their superior performances. Good finishing foremen do not just happen, though; these owner-managers selectively pick the foreman candidate, then deliberately and carefully train him to make decisions and to supervise employees according to standards and methods used by the owner-manager. Finally, they back the foreman's decisions with the workers, while remaining open to employee grievances. Shutting off grievances invites unionization.

One factor contributing to employee unrest following the appointment of a foreman are the feelings of resentment and jealousy among those workers not chosen. These feelings may be especially strong in a newer company where some of the original employees hired are still working in the production group. Often, the original group developed a special allegiance to the owner-manager as they collectively struggled for survival. When enough success and growth is achieved to warrant the appointment of a foreman, there may be some reluctance to accept direction from anyone other than the original founder. Faced with such resentment, foremen must be skillful in human relations in order to avoid serious difficulties.

Marketing

The success of an independent finishing firm depends mainly upon the degree of activity of its customers. No matter how efficient and well-managed a job shop may be, it cannot survive unless its customers generally enjoy successful operations. This dependency on the fortunes of others presents a problem of maintaining a sustained high volume of activity.

The problem may be substantially reduced if firms build an aggressive marketing effort and a variety of customers, rather than depending on just a few. Operating activity may be reasonably stable at a fairly high level throughout the year if a company can sell to customers that, as a group, have complimentary seasonal requirements. Some shops, however, sell a very high proportion of their output to only a few customers in one or two industries and have, therefore, become so tightly tied to specialized customer requirements that they have lost much of their flexibility. Although answers varied widely, the questionnaire results and field interviews indicated that it is not uncommon to find finishing firms having 30 to 40 percent of their sales going to one customer, and 50 to 60 percent of their sales confined to one industry. In a few cases, as much as 90 percent went to one industry. Such firms often find it difficult to adjust when decreases in the demand for their customers' products reduces the demand for their finishing services, thereby creating excess capacity. Too often the response of such firms is cut-throat pricing, in order to capture finishing work being performed by other independent shops at more reasonable prices.

Few firms surveyed are taking effective action to ease this dependence on only a few customers. Commonly, a shop owner acts as his own part-time salesman, waiting for jobs to appear or for prospective customers to publicize their needs. One factor encouraging this policy is that it is difficult for the smaller firm to efficiently utilize a full-time salesman; one or two large orders might tie up a three to six employee shop, leaving the salesman with little to do while waiting for unused capacity. The possibility of firms sharing a salesman or using a manufacturers' representative was rejected by some as not being feasible. However, several survey companies are using manufacturers' representatives to achieve a stable and significant sales volume from a diversified customer base. In fact, 26 percent of questionnaire respondents are using a manufacturers' representative. Of equal importance to balancing the workload, a company is able to be more selective in choosing which of several jobs it will bid on. Profits can be enhanced by avoiding customers and jobs offering little profit potential.

The marketing program, in other words, should be more than simply "waiting for the phone to ring". This, to our astonishment, is the way many firms described their marketing effort. Filling orders by telephone queries often means responding to bargain hunters. Higher profit companies fill their capacity with orders in markets they have aggressively sought out. Such customers provide rates giving a more attractive return on investment.

The field study also revealed instances where independent finishers made a practice of maintaining some idle facilities to enable them to retain the flexibility to reject low profit jobs and accept the high profit jobs, such as "rush" orders where the customer is willing to pay a high price because of the time schedule required. Although we discuss this more fully in the chapters on pricing, we point out here that this is sound marketing strategy only if the total contribution margin[2] on the "high profit" jobs ultimately exceeds the total contribution margin on the "low profit" jobs rejected. Of course demand for the finisher's services must be such as to make this possible. If such idle capacity becomes permanent because of low demand, acceptance of some or all of the low profit jobs, as long as they contribute to profits, might be the best decision. More on this later.

In sum, first, professional marketing efforts need to be increased by most firms, whether by more sales emphasis by the owner-manager, or an addition of a sales person to the management team, or by manufacturers' representatives. Second, as a general rule, such efforts should be directed at obtaining a diversity of customers in different industries whenever possible.

Technical Considerations

Technical proficiency in the finishing industry has three support columns — chemical, mechanical, and metallurgical engineering. Inadequate skills or performances in any one can result, and often does, in achievements certainly less than the theoretical peak, and, at times, even less than what is needed for acceptable results. Few companies last long in the finishing industry when they lack a thorough understanding of chemistry fundamentals related to finishing operations. There is some art in every science, but solid metallurgical expertise needs to be combined with a sound understanding of chemistry. The finishing production process needs to be approached more as a science than an art. In addition, production-oriented finishing companies require mechanical

2. Contribution margin is equal to sales minus variable expenses (see Chapter IV for elaboration).

engineering know-how, necessary for the design of material handling mechanisms and for tooling requirements, such as racks and special anodes.

The successful companies that we surveyed demonstrate above average ability in applying these engineering disciplines to (1) adapt their traditional processes to meet, in an economical way, higher quality or performance standards made possible by a competitor's new technological advance; and (2) design scientifically, rather than empirically, the process factors such as equipment, tooling, chemicals, and finishing methods, and then to control these elements for lowest costs meeting acceptable uniformity and quality standards.

Worker Motivation and Productivity

As in any organization composed of humans, personnel problems rank high on the list of challenges facing independent company managers. Hiring experienced and competent employees is of constant concern. The problem is even more severe, however, in industries made up of relatively small firms, such as is true of the metal finishing industry. The fact that the lack of experienced and competent help was listed as the second most important problem both for individual firms and for the metal finishing industry in general (see Chapter I) attests to its importance here. Small firms are at a relative disadvantage in conducting training programs, in providing the job security that workers want, and in competing with the wages and fringe benefits of larger product-oriented companies.

Labor turnover in job shops fluctuates with economic conditions. When depressed economic conditions make jobs hard to find, job shop labor turnover is relatively low. When the business sector is operating at a high level of employment and output, as it did in 1973 and 1974, job shop turnover increases significantly. During such periods, marginal workers must be hired, bringing problems of higher absenteeism and turnover, with lower output and quality of work performed. One company needing an average of 42 workers throughout 1974 — an excellent sales year for the finishing industry — had to employ a total of 120 different workers to sustain a level work force at 42. Not only did workers with long company tenure leave to take higher paying jobs in large manufacturing concerns, but many of the marginal workers available in the work force at that time turned over very quickly after being hired, unwilling to accept the rigors demanded by a permanent job. Frequently, they worked less than a month before quitting. By contrast, during 1976, a year of moderate economic activity, the company's turnover was considerably lower, requiring only 60 workers throughout the year to staff the average work force of 37 men. Obviously, the company's labor productivity suffered markedly in 1974

Labor problems frequently are influenced by whether the plant is an automated or a manual operation. Working conditions in the former are relatively agreeable. Moreover, fixed costs (equipment depreciation expenses) are substantial, so managers of these firms place more emphasis on employee retention and satisfaction in order to keep equipment operating at a high level of activity.

Employee problems are more conventional in automated companies, generally similar to those of comparable industrialized operations. Examples of their typical problems are union negotiations on wage rates, fringe benefits, and shifting of work assignments as the workload changes throughout the day. The latter problem has been solved in some companies by agreeing to pay a worker the highest rate earned throughout the day regardless of the number of types of jobs performed. For example, a truck driver may rack or buff parts for two hours while waiting for a delivery and be paid at the higher truck driver rate rather than the lower rates of the racker or buffer.

Labor problems in the nonautomated finishing plant are influenced heavily by the physical environment in which the employees work. In contrast to the automated plant, a small manual shop's floor is often wet from finishing solutions splashed or dragged from the tanks. Workers typically require protective footwear, frequently knee-high boots. Additionally, the nonautomated process normally requires lowering, lifting, and transferring parts into, out of, and between plating and rinse tanks, barrels, and buckets. For parts or racks that are heavy or bulky and require several operations, the physical effort involved is taxing.

Although job shop managers recognize the benefits of minimizing the inconveniences of the job and workplace, only so much can be done without changing to automated equipment, where material transfer is mechanized and plating and rinse solution spills are more easily controlled. Managers report that, despite their own efforts to better the work environment, workers — particularly the new and younger ones — frequently become dissatisfied and quit, either to take employment elsewhere when the economy is healthy, or to draw unemployment compensation, when possible.

Companies confronted with labor turnover seem to be better off located in urban areas since there is a greater likelihood of an untapped supply of workers. In such areas, companies are able to pay wage rates somewhat below those paid by companies offering similar services in rural areas, where the labor supply often is limited. Finding it difficult to rehire or to recruit replacements, the latter companies are less flexible in furloughing workers during temporary business lulls.

Not only must managers of some rural companies pay higher rates and staff maintenance costs, but fringe benefit costs are likely to be greater for this more permanent group of employees. In urban shops where tenure may be short, employees may not work long enough to qualify for many fringe benefits, such as insurance, vacations, and pensions. Because finishing is labor intensive in nonautomated plants, a company's location and access to a supply of available workers are important determinants of success, particularly where demands in the finishing marketplace fluctuate.

A word of caution is necessary regarding the last two paragraphs. Sometimes the increased worker stability and efficiency in the rural areas more than counterbalance the lower pay and fringe benefits in the urban areas. Also, with increased labor turnover in urban areas comes the additional cost of hiring and training new employees. In sum, each firm must consider the area in which it operates, in order to make a final determination of whether a rural or urban location best fits its needs.

How can such problems be solved, and how are they being solved by the successful companies? Foremen in the industry must become a more influential force in worker motivation, as they frequently are in the more successful firms. This is a serious challenge in manual finishing shops where the work can be physically demanding in the face of hot, humid, and generally unpleasant physical conditions.

The lack of effective employee motivation is evidenced primarily by the extremely low productivity gains in the industry over the past three decades. While increased mechanization and automation gradually may find their way into some industry sectors — having a positive effect on productivity, certain types of finishing, especially low volume work, may never be automated. Productivity gains in those cases will be vitally dependent upon individual workers and upon management's ability to motivate them.

Besides the demonstrated productivity gains possible through improved relations among production workers, foremen, and owner-managers, two related management practices of successful companies are worthy of mention. Some managers in finishing-related operations are recognizing that they have no monopoly on ingenuity, and that valuable ideas and advice can arise from production line workers under proper conditions.[3] Recent experience has shown that production workers can be a rich source of profitable improvements when suggestions systems are tied to incentive plans. Even organized labor, which often assumes a negative stance toward incentive plans, has reacted favorably to certain worker participation-incentive programs.

Incentive programs in the finishing industry are generally of two types: individual worker incentives for the manual finishing shop, and group incentives in the mechanized production-oriented shop. A

3. Successful examples are Donnelly Mirror Co. and Harman Industries (both having extensive finishing operations).

worthwhile goal of incentive plans should be to enlist the brain power of individual finishers in devising methods to decrease idle tank or equipment time—idleness while waiting for the next job or operation. Without worker participation and involvement, these bottlenecks must wait resolution by the foreman or the owner. Experience has demonstrated that worker participation-incentive systems effectively encourage finishers to think of new ways to minimize the number of delays and to shorten them when they do occur. It has been demonstrated that everyone benefits when production employees work more effectively and efficiently.

A major obstacle to incentive systems for finishing firms has been the challenge of recording time data for production operations. Successful companies do have reliable job cost accounting systems, however, and most firms could have. The potential for electronic data entry and computerized systems, moreover, suggests that shop floor data recording problems can be greatly minimized, thereby making standard-setting for incentive programs less of a problem. Admittedly, without reliable cost data, standards and incentive programs will not work.

Production-Related Factors

While an in-depth analysis of differences in production operations between successful and struggling finishers would disclose many factors, three major ones came to our attention. Orderly plant layout and relatively smooth work flow characterize most successful finishers. By contrast, many struggling companies operate in cramped, chopped-up, dark buildings. Such inefficient facilities spring from the formation of many companies. Typically, a venturous technician starts in business by renting the smallest, least expensive building he can find. To be near the customers he has targeted, he often selects a site in a congested industrial section, having little or no surrounding vacant land. Never quite certain that business will prosper long enough to justify building a new plant meeting the special needs of a finishing operation, periodic expansions are made by adding a module to the first building or by setting up satellite lines in other nearby buildings. The result often is a plant layout nightmare.

Through proper foresight, or just plain good fortune, successful companies are more likely to have selected buildings affording adequate space for expansion and for the efficient organization of equipment, properly distributed to permit wide aisles and sufficient storage areas near tanks and machines. Space-cramped companies, on the other hand, have parts stored in bins or boxes several feet high, which are often waiting in narrow, blocked aisles to be moved to the next operation. Such inconveniences frequently require that finishers crawl around or step over boxes and bins to find parts to assure that all of the batch has been located. Resulting inefficiences in the production process create penalties estimated as high as three to six percent of total costs.

Lower chemical costs are another attribute of successful companies. For the industry as a whole, total material and labor costs frequently comprise almost two-thirds of all operational costs. While many survey companies were attentive to their readily observable labor inefficiences, only the more successful ones employed procedures and practices adequate to minimize expensive material costs. Line workers often had the authority to remove chemicals, metals, or buffing materials from the storeroom and enter them into the process. In the struggling companies we did not observe much control over these costly items. Results such as the following were not uncommon:

1. Finishing-line technicians intuitively modifying solutions in attempting to recoup quality, when the problem is of an electromechanical rather than chemical nature.
2. Poor finishing quality caused by unmonitored solutions varying from acceptable limits — in severe cases, finishing lines were shutdown while solutions were analyzed and treated in order to return them to acceptable limits.
3. Entire solutions erroneously dumped by finishing technicians in a blind search for better results. Analytical procedures sometimes demonstrate that minor chemical additions could have solved the problem.
4. Finishing technicians dutifully making routine chemical additions to solutions, without prior solution analysis or subsequent analysis of the addition effects. Subsequent analysis of past routine chemical treatment procedures sometimes demonstrate that daily solution treatments, without prior and post solution analysis, result in wasteful and costly additions.
5. Subsequent analysis of discarded polishing and buffing wheels demonstrate that they were indiscriminately or prematurely changed, thereby causing increased costs.

Larger companies generally combat these inefficiencies by adding a qualified chemist and instituting strict and formalized process analysis procedures and records. The issuance of materials from stores and their insertion into the processes are also controlled. Smaller companies often are unable to justify the cost of a full-time chemist; they are forced to rely on the "intuition" of finishing technicians to carry out chemist-dependent activities. The result is seldom optimal.

The value of a skilled chemist in material cost control was demonstrated in one company: when the chemist was away on vacations, chemical costs regularly jumped 20 to 25 percent. After several occurrences, the company instituted an intensive training program for the line personnel responsible for maintaining the chemical solutions. Despite these efforts, cost increases due to the absence of the chemist never were less than 10 percent.

Still, the smaller companies face a dilemma since it is hard for them to justify a full-time chemist. Regardless of how they do it, these companies must set up a formal chemical monitoring program. Chemical suppliers have customer application engineers to assist in setting standard levels, in determining monitoring procedures, and in training required personnel. Small companies should be entitled to the same help the suppliers furnish the large captive and independent shops. However, because they lack the leverage enjoyed by a large purchaser, the small shop may have to settle for less. Continuing education programs are one alternative. Another option is for a number of firms to arrange to share a chemist.

As chemical costs continue to increase, the differences in process-chemical costs between the knowledgeable and the "cut-and-try" companies will rise. Sooner or later, the struggling company will have to plug this cost drain or go out of business.

Finally, the value of a qualified chemist will increase along with the increased application of pollution abatement technology and equipment. The only hope for deriving an economic return from the pollution control investment is recycling chemicals and metals. Fundamental knowledge of chemistry will be essential in the operation and the monitoring of sophisticated pollution control apparatus; it will also be vital in adjusting process variables to accommodate recycled chemicals that may not always agree with the analysis specifications guaranteed by suppliers of virgin chemicals.

Less efficient utilization of plant and equipment is a third major production-related difference between successful and struggling finishing companies. Inefficient utilization of capital facilities is costly for labor-intensive manual finishers; it is devastating to mechanized companies. Fixed depreciation charges can be a substantial proportion of total operational costs. Although we cover fixed charges in detail in later chapters, a simple example will suffice here. When amortized over a ten-year period, the depreciation charges for a machine costing $300,000 amount to approximately $115 per working day, regardless of whether the machine is used for one, two, or three shifts. The less the machine is used, therefore, the greater the average cost of production per piece. Capital equipment scheduling decisions, therefore, do have a substantive effect on the final profit or loss figure.

A number of survey companies operate their plant and equipment only one shift. Successful companies, recognizing the drain on profits caused by idle finishing equipment, schedule two and three shifts. When business and workers are available, they operate on Saturdays and Sundays as well, spreading depreciation expenses and other fixed charges (insurance, taxes, etc.) over maximum operating hours. They find it pays to overlap shifts and to have maintenance men on call at all times for maximum machine uptime. Some even have pools of housewives available who will, on short notice, come into the plant to substitute for workers unable to work their regular or extra shift.

Obviously, in order to operate on a multi-shift basis more finishing jobs are needed, requiring even more sales effort to fill the idle capacity. Many companies obtain enough business for single-shift operation with little outside customer contact. Tripling their sales levels for a three-shift operation would certainly require adding field sales personnel.

Many successful companies point out that their around-the-clock operation actually attracts sales they otherwise might not get. Some highly motivated service finishers pick up customer's priority work after their first shift, finish it during the night, and return it to the customer before his first shift on the following day. This quick reaction capability has enticed many customers to send all their work to a finisher with this capability — knowing that in an emergency they can depend on overnight work.

Besides placing increased demands upon the sales function, around-the-clock operation demands that the second- and third- shift foremen be highly qualified. Commonly, they alone are responsible for the entire plant after the owner-manager leaves for the day. Thus, decisions over the entire range of operational matters by the second- and third-shift foremen can go a long way in determining the firm's profits or losses.

To permit this first step towards decentralized management, daily operational reports showing direct expense ratios, reject levels, chemical and metal costs, etc. are mandatory. Captive shops routinely generate and use many such reports on a daily basis. As the independent companies acquire computers equipped with electronic data entry from the production floor, these managers also will be able to review each morning the reports of the previous day's and night's operations.

Using Accounting Information Effectively

The root of many management problems in the industry is the aversion of the manager to administrative tasks, particularly the effective and efficient utilization of accounting information for decision-making purposes. Although considerable attention is devoted in later chapters to accounting and its uses, a few general comments are in order here.

Some managers view with outward contempt even minimal record-keeping chores. A few survey companies did not even prepare a profit and loss statement. In those companies, the requested survey data came from the owner-manager's personal tax return. Finishers tend to be contented, and often exhilarated, working around finishing tanks. They frequently are bored and miserable by "desk" work. Successful operations, however, require balanced management attention. If the owner-manager is unwilling or unable to perform such functions, he must make use of the services of those who are. When a manager's aversion to administration is so dominating that it prevents even minimal record keeping, his company is doomed to remaining small and relatively unprofitable at best, to failure at worst.

Some managers rationalize or excuse their administrative deficiencies in a number of ways. One common way is to cite the difficulties of adequate data accumulation. Record keeping, they say, is difficult in the finishing room, where many small-value jobs may be processed by a group of workers rather than by individual craftsmen. They feel it is economically impractical to have the workers, for example, record on job tickets the time spent on individual operations. They further argue that there is no set routing for individual jobs through their shop because specification requirements often differ, making record keeping difficult if not impossible.

Data accumulation isn't easy, of course. Nevertheless, successful companies demonstrate that workers can and do record basic operational job cost data when management demands it, and when workers are provided with the proper forms and understanding regarding their completion. It is important to cite the critical need for the information to foremen and workers, and to indicate how essential it is to successful operations — and hence to the future of their jobs.

Data accumulation is not the real problem, we feel. The real problem is one or a combination of three subproblems. First, many managers still feel they can operate by the "seat of their pants," that accounting information is a necessary evil because of tax or other governmental regulations, but that it is useless as a helpmate to them in running their businesses. Second, some managers who understand the importance of accounting information do not adequately understand the accounting information provided to them, nor do they know how to interpret and utilize it in the decision-making process. Third, in many cases managers are not receiving sufficient, or the best, accounting information.

These problems are considered in detail later. Suffice it to say here that successful managers are the ones who are utilizing their accounting systems to provide information (such as job costs) on a timely basis, permitting quick corrective action where needed; financial data useful in obtaining capital as needed; ratio and other analyses to determine trends in their companies' operations; data indicating poor customers and unprofitable jobs; and data for effective pricing, among others.

Management Recruitment

Another serious problem is attracting young, dynamic people into the industry. Though difficult, more effort is warranted in this regard than has been exerted in the past. Any number of small- and medium-sized business schools throughout the country are having difficulty in placing their graduates. It seems to us the industry should make a greater attempt at recruiting some of these students. A number of approaches can be taken. First, the National Association of Metal Finishers might act as a clearing house and take on central recruiting functions. For example, the National Association could develop attractive informational brochures about the finishing industry and distribute these to placement personnel at selected schools throughout the country for use by students seeking employment. Interested students could contact the national office and be placed in contact with someone in the local area.

Second, selected members of the Association in various areas throughout the country could assume the responsibility of determining the needs of Association members in the area, i e., job descriptions, starting salary figure, benefits, and future prospects. Armed with this information, these people could recruit on the campuses of colleges and universities in the area. Many such schools have courses in "small business," and professors teaching these courses can be contacted to start the ball rolling. A surprising number of business students are interested in working for small business, but they have little opportunity because they are not actively recruited and are unaware of the opportunities. Keep in mind that the placement personnel and the professors at these schools have a vested interest in seeing to it that all the graduates of the school obtain employment. Given adequate data, most of them will do a good job in making the student aware of the opportunities in the metal finishing industry, in establishing recruiting schedules, and making space available to the recruiter when he visits the campus.

Framework for a Successful Company — An Example

We surveyed a number of successful companies. From that group we have selected one to describe, in order to reflect as a total picture the way one company has fulfilled the more important criteria for success. Obviously, success might be achieved by other companies using different specific techniques and procedures; but the example might aid in seeing and understanding the interrelationships and interdependencies of major functions which must be performed well if success is to follow.

A key factor influencing the success of one exemplary company in our survey was a management team composed of second- and third-generation owner-managers, a father and his son. Besides the combined years of experience the two principals brought to the company's daily operations, their full-time involvement made possible a division of labor that fostered advantages inherent in specialization. The team approach allowed each manager to devote his full energies to running a two- or three-shift operation while occasionally taking well-earned vacations with the assurance that the other was managing the company during his absence in accordance with demonstrated policies and procedures.

Of critical importance was the two managers' unique understanding of the chemical, mechanical, and metallurgical fundamentals for successful metal finishing. They excelled in all three disciplines, in addition to being skillful managers. As one example of superior mechanical know-how, the managers designed and produced their own racks and other tooling. Cost savings resulted for the comapny for the following reasons:

1. Particular attention was given to rack and polishing fixtures in order to facilitate easy part loading and unloading. Designs stressed security to prevent parts from falling into solutions.
2. Specialized rack and fixture setups often were imaginatively assembled from generalized and reusable components.
3. Special rack designs or attachments minimized the need for unique and costly conforming anodes.

Although the company experienced minor direct savings over purchasing commercially available racks, indirect economies and payoffs were far greater. An in-house bank of electrochemical and mechanical expertise — specifically applicable to the operations of this company — was developed over the years and is tapped each time a new job is received for process design. Tailored designs are more apt to approach maximum quality and production rates, as compared to the compromises that often are necessary when using off-the-shelf universal tooling. Also, having an in-house staff meant that rack, buffing, and polishing setups were constantly scrutinized for maximum efficiency.

As an example of superior tooling know-how, the company designed imaginative rack stripping systems that are integrated into the automated plating lines. Whereas many companies delay stripping their racks for as long as possible, resulting in heavy concentrations of metal build-up, this company strips the racks on every cycle by this integrated process — and in a fraction of the time commonly required. The benefits are many. Damage to the racks is reduced since the time the rack is exposed to the stripping acid is shortened. Rack efficiency is also enhanced; a clean rack is closer to being physically identical each time it goes through the plating cycle, thereby removing one more variable that otherwise might adversely affect quality. The company also finds that rackers are able to obtain 10 to 12 percent greater productivity when working on a line with clean racks. Too often rackers are delayed and frustrated when trying to load parts on racks having excessive buildup.

Having an in-house tooling design and manufacturing capability is economically feasible for only the largest of the companies in the industry. An operation much smaller than this 125 employee company would find it a challenge to keep the rack and tooling design personnel profitably occupied and also to employ economically the rack-room production equipment. Nevertheless, because this company was so obviously successful in exploiting its rack tooling know-how, it suggests the potential payoffs inherent in increasing the average scale of operations throughout industry.

The manager's expert electrochemical knowledge sometimes permitted the modification of conventional processes to new higher standards made possible by a competitor's advanced but more expensive process innovation. For example, through experimentation with the conventional cyanide-zinc plating process, the company was able to achieve the same high brightness level otherwise attainable only through the more expensive chloride-zinc process.

A vital policy resulted in this company's being one of the few in the industry utilizing a marketing staff composed of several professionals. The successful penetration of new markets and customers reduced the company's dependence on a few customers or a single industry.

Owing to an active sales force and certain other factors, the radius of their market was about 50 miles larger than that enjoyed by the average company. Finishing customers as far as 150 miles from the plant made up their market. Higher shipping costs in the fringe areas were overcome by extensive use of automation, by material discounts resulting from buying in large quantities, and by demonstrated production economies inherent in greater overall efficiency and multi-shift operations. The aggressive sales effort, besides dramatically enlarging and diversifying the markets, made shifts from declining customers or markets a profitable reality.

Because the company was one of the largest purchasers of buffing and plating materials, it successfully made persistent demands on its vendors for technical assistance. The owner agreed to concentrate his material purchases with a particular vendor as long as the supplier sold at competitive prices and provided extended in-plant technical support. This advice included setting standards for baths, aiding in the monitoring of solution chemical levels and plating quality, and assisting in process design and chemical content determination for new finishes and plating methods. Chemical refinements were constantly reviewed, resulting in cost reductions.

The company, through quantity discounts, saved 10 to 15 percent on materials, and it enjoyed valuable expert advice on chemical usage without hiring highly priced technical workers. One full-time chemist handled all process control and pollution abatement matters.

In pursuing its goal of diversification, the company sought customers in new industries whose historical sales patterns complemented the peaks and valleys of existing customers. The company diversified its customer base to include building hardware, automotive, appliances (both residential and commercial) and food marketing equipment. An impressive result of this policy was evidenced in the ten straight years of annual sales improvements — even during the general economic downswing of 1974-75.

Consistent with the marketing diversification policy, the company sought a balanced complement of finishing equipment with respect to size capacity. Approximately fifteen years ago, in the face of accelerating labor costs, the company decided to phase out all hand finishing and concentrate on automated processes. The automation was selected so that the company would still be competitive on jobs with as few as 100 pieces. Individual finishing lines were designed to cover specific ranges of part dimensions, and the full range provided an overall capability adequate to meet the majority of its market's needs. When the backlog for a particular line shrank, the sales force concentrated its efforts on finding finishing orders to fill this void.

Over the past decade, the company acquired at least one major piece of equipment each year. In three-fourths of these cases, they selectively purchased an automated line at a used equipment auction at an acquisition cost approximating 20 to 25 percent of the original price. By constantly following the used machinery auctions, the company could be extremely selective and actually found machinery only three to four years old at substantial savings. In nearly every case, the equipment was refurbished by the maintenance tradesmen and then installed at a total cost of about one-third of the price of new equipment. The owner's superb mechanical knowledge was a key factor in making these ventures successful. Impressive equipment performance standards attested to the management's skillful evaluation and selection of the used equipment, and to the thoroughness of the maintenance mechanics in refurbishing and installing it.

The company's foremen also were assigned vital roles in achieving maximum utilization of the equipment. First, the foremen were trained thoroughly in the equipment's design and operation. Machine components that were wear-prone received special attention. It was the foreman's responsibility to monitor the entire line under his supervision. Maintenance workers taught the foremen how to sense developing problems before a crisis occurred, so they could be attended to before breakdown. Every means available was exercised to keep the line running until a nonscheduled period or a weekend, when major corrective maintenance could be performed without disrupting a productive work schedule.

Special emphasis was given to training and motivating the foremen so they were in fact an extension of the company owners. Each foreman received production schedules and specifications and was then held accountable for the productivity and quality standards of his respective line. Problems and transactions with support activities, such as maintenance, rack-room, the chemist, transportation, etc., were handled independently of the owners, unless solutions required actions or decisions reserved for them.

The owner-managers enjoyed direct contact with operative personnel and plant equipment and, therefore, relied heavily on personal contacts and visual inspections to evaluate performances. In addition to on-the-spot monitoring, formal reports on production rates, chemical analyses, and reject rates were studied daily. Ratios of various expenses to total costs were analyzed weekly; these expenses included direct labor, various buffing and plating materials, utilities, and transportation. Because the two managers closely monitored daily production operations and records, the weekly expense and financial reports seldom presented surprises — no significant deviations from trends they observed to be developing during the week.

In evaluating what comprises the framework for a successful finishing operation, an early conclusion was that a few high-visibility functions performed to superior standards will not alone produce high profits. Controlling overhead expenses, especially those of a nonproductive nature, is a critical but often overlooked factor. One striking difference between this highly successful company and the "average company" shows up in clerical expenses. In contrast to this company, which used one clerical person for each $1.5 million of sales, the vast majority of finishing companies surveyed employed approximately one clerical worker for about each $0.5 million of sales. Several factors accounted for this company's more favorable ratio.

1. A minicomputer processed payrolls, accounts payable, accounts receivable, and job cost records. Simple but effective computer programs were developed to create timely management reports so that, when they were combined with daily contacts and the production, process control, and quality control reports, both managers were thoroughly informed of all operational details. Excellent management control was possible, and it was exercised.
2. Because the well-trained, experienced, and highly motivated sales force successfully met sales goals and redirections, the managers devoted minimal time to field sales calls, thus freeing them for in-plant matters. It was the direct sensing of conditions during

"hands-on" operation that reduced the need for extra administrative reports, normally required by "absent" managers for control.

The profit position of the company was also protected by guarding against submitting faulty quotes in response to customer bid requests having difficult requirements. These included suspected problems that might arise from extreme tolerance requirements or tight finish uniformity demands relative to complex part configurations, a new finishing technique, etc. Three policies minimized the chances for loss of bids that otherwise would be risky:

1. In areas of extreme uncertainty, the company, enjoying above-average customer confidence, offered to finish the parts on an actual cost plus fixed fee basis. Because of a long and successful record with its customers, many low-volume jobs were handled this way. On higher-volume jobs a small batch was made available by the customer for a trial run with actual costs to be billed, followed by negotiations for the price of the remaining lot.
2. A second method, preferred by the customers, is a variation of the first; the customer provides a sample batch of the parts which permits the company to analyze the costs under actual plating conditions. A competitive bid for the entire job is then submitted. The company also likes this procedure because it allows for process experimentation which sometimes permits it to negotiate minor design changes with the customer, and to test out possible efficiencies that competitors may overlook.
3. The third policy is perhaps most important of all; the company, when it is unable to price a puzzling bid request confidently, either refuses to bid or estimates its cost on a "worst-case" situation, then adds a contingency safety factor.

Profits in this company are, therefore, affected not only by the quality of the job it receives but also by the jobs the company chooses to avoid. To enjoy the luxury of this policy, the company has ample sales opportunities coming in from its field sales force — highlighting again the importance of aggressive marketing.

The net result of these factors and other less obvious elements is that the company enjoyed a measure of success achieved by only a few in the industry.

Concluding Remarks

Most of the successful firms we surveyed were members of the National Association of Metal Finishers. Partially this is true because most of the larger firms are members, having an overwhelming portion of total employees, total capital investment, and total sales of the industry. Since only 18 percent of all independent metal finishing firms are members of the Association, few of the smaller firms have affiliated with it.

Small firms, however, can benefit by membership in the Association as much as and possibly more than large firms. Smaller firms are less able to afford accountants, lawyers, engineers, management consultants, and others who can aid them in operating efficiently. Without such help, these firms are more likely to establish pricing policies that will be detrimental in the long run to their goal of profit maximization. Their cost information is more likely to be sketchy or nonexistent, thus denying managers a useful tool in producing and distributing their products and services efficiently; this in turn keeps profits from rising to the maximum level possible.

Successful firms are more likely to be members of the Association, therefore, because the educational programs, committee meetings, and other activities of the Association aid in making them successful. Consequently, it would be in the best interests of these smaller firms if they would join the Association and take an active interest in its activities. It would help them keep abreast of current developments in their industry and, more importantly, improve their long-range profit potential.

The National Association of Metal Finishers has grown rapidly in size and in the development of programs and information for its members during its twenty-one years of existence, particularly during the last ten years. However, more effort is needed in two areas in particular.

First, even greater effort should be placed on its function of conducting membership drives, if many of the problems of the industry are to be solved. The member firms, through personal contact and armed with informative brochures, should make concerted efforts toward getting these nonmembers, particularly the smaller firms, into the Association.

Second, even more seminars and workshops, related to specific industry and firm problems, should be developed and presented for the benefit of members and, maybe even more importantly, nonmembers of the Association. Any number of possibilities exist, but the Association might concentrate initially on making programs available on understanding and using financial and management accounting data in general; and on specific areas such as cost analysis and control; pricing policies; and the acquisition, training, and maintaining of adequate, competent personnel at all levels.

Any number of good continuing education programs exist outside the Association. Varying in length and complexity, they cover an untold variety of topics. By nature these are more general in their coverage because of the various backgrounds of the participants. Certainly, finishers should take advantage of these programs, but the Association should not utilize its resources to compete unnecessarily with such programs. The Association's effort should be directed at developing educational programs designed exclusively for the metal finishing industry, based on its problems, and using its data in examples, problems, and solutions presented.

In any case, if the success of the industry in general and specific firms in particular is to be enhanced, proper attitudes must be developed. People in the industry must be convinced that continuing education is necessary and must want to do something about it. They should be ready to share information and experiences with other participants in meetings and seminars. This does not mean they ought or need to divulge confidential information. It does mean participating freely in discussions on problems common to the industry, being willing to seek out and attend seminars, and engaging in independent reading and study. Perhaps most important — and most difficult — is for each person to determine objectively his management weaknesses, acknowledge them, and then to take action to correct them. Until this is done, help is not possible.

III

ACCOUNTING FOR METAL FINISHING FIRMS A MANAGERIAL EMPHASIS

Our survey indicated that a significant number of metal finishing firms' managers fell into one of three categories:

1) Those who felt accounting, except as required for filing tax returns, was unnecessary and useless;
2) Those who felt accounting was necessary, but didn't understand it well enough; and
3) Those who felt it was necessary and had a basic understanding of accounting, but were not being provided with the best or most useful information for operating their businesses.

One of the major purposes of this monograph, is to show why accounting data are necessary and useful, and to indicate which types of accounting data are most useful for managers of independent metal finishing firms. We hope to convey enough understanding of accounting so that managers can first ask the right questions of the accountants they employ and second, understand the outputs of accounting well enough to make knowledgeable decisions regarding their firms.

Some businessmen, particularly those in the small firms, feel that accounting is of little use to them. After all, they reason, their association with the operations of the business is so close and continuous that accounting reports and statements can add little to their knowledge. Such an attitude is wrong and dangerous for a number of reasons. First, in even the smallest of business ventures, it is unlikely that the manager can retain all the financial facts of his business "in his head," while at the same time being involved in the myriad of day-to-day operating decisions which consume his time. Second, and more importantly, unless he is reasonably well-versed in accounting, the finishing firm manager is unlikely to see the interrelationships and interdependencies between the data provided by accounting. These data, in fact, can tell him a number of stories about his business—where it has been, where it is headed, how its course might be altered for the better—stories he must understand if he is to operate successfully.

The problems of operating a small business successfully are no different, generally, than those of large business—less complex, of course, but essentially similar. Pricing decisions must be made, costs must be controlled, financial strength and stability must be sustained. All these problems require accounting data, properly interpreted and understood, for satisfactory solution. Our survey indicated that this attitude of not needing accounting data, particularly on the part of the managers of the smaller firms, resulted directly in a number of bad practices, such as cut-throat pricing. Such pricing was not deliberate in most cases. It resulted because the manager was not knowledgeable concerning the cost structure of his firm and/or had little understanding of pricing principles or the effect that unsound pricing decisions would have on his firm or the industry.

In many cases, those receiving accounting reports received nothing more than periodic income statements and balance sheets, often without interpretive comments and analyses. These statements, as we shall see, can be highly useful, when properly understood and interpreted, for certain purposes; but much more is needed. A major purpose of accounting is to make use of data relative to past operations and present conditions in order to plan for the future, as well as to prepare and present data which will aid in the decision-making process, i.e., in choosing from among alternative courses of action. Income statements and balance sheets can act as part of the foundation data for doing these things; but budgets, cost reports, forecasts, and the like also are necessary. All decisions are futuristic by nature, since the past cannot be altered. Planning needs to be utilized much more in the metal finishing industry than presently is the the case. Suffice it to say, no firm can expect to operate successfully over an extended period of time without sufficient and appropriate accounting data.

For the remainder of this chapter we will discuss the balance sheet, the income statement, the funds statement, and budgets to gain a basic understanding of what they are, their strengths and weaknesses, and the stories about the firm they tell. In Chapter V we will discuss financial statement analysis and how accounting reports can be utilized in certain aspects of the decision-making process. Other uses are discussed in Chapters IV and VII.

Balance Sheets

A balance sheet reflects the financial position of a firm *at a point in time*. If a balance sheet is prepared at December 31, 19____, for example, it literally is an attempt to reflect the condition of the firm exactly at the stroke of midnight on that date. It is like a snapshot picture of the firm, as though all the operations came to a standstill at that instant in time, and the financial status of the firm determined. Following is an example of a typical balance sheet (excluding much of the detail which would appear):

Three major factors are critical in the determination of financial condition: assets, liabilities, and stockholders' (owners') equity.

Assets

An asset is anything objectively measurable in dollar terms which has anticipated future service utility to the firm. Assets are the "things" utilized to generate profits. Profits must be earned; efforts must be

METAL FINISHING FIRM
BALANCE SHEET
As of December 31, 19—

ASSETS

Current Assets		
Cash	$ 7,000	
Accounts receivable (net of allowance for bad debts of $4,000)	35,000	
Inventories	11,000	
Prepaid expenses	5,000	
Total current assets		$ 58,000
Plant, Property and Equipment		
Land	$ 20,000	
Buildings and improvements (less accumulated depreciation of $40,000)	220,000	
Machinery and equipment (less accumulated depreciation of $50,000)	70,000	
Total property, plant and equipment		310,000
Other Assets		
Goodwill		5,000
Total assets		$373,000

METAL FINISHING FIRM *(Continued)*
BALANCE SHEET
As of December 31, 19—

LIABILITIES AND STOCKHOLDERS' EQUITY

Current Liabilities		
Accounts payable		$ 15,000
Accrued payroll payable		2,000
Social security taxes payable		800
Income taxes payable		12,000
Current portion of long-term debt		10,000
Total current liabilities		$ 39,800
Long-Term Liabilities		
Mortgage payable	$100,000	
Due to officers	20,000	
Total long-term liabilities		120,000
Total liabilities		$159,800
Stockholders' Equity		
Common stock	$160,000	
Retained earnings	53,200	
Total stockholders' equity		213,200
Total liabilities and stockholders' equity		$373,000

generated to attain them. Assets measure the services available for future use (from the balance sheet date) for such profit generation. This "future service utility" is the thread of similarity which makes all assets similar; current assets are similar to noncurrent in this regard, cash is similar to land, buildings similar to goodwill, etc. In this sense, also, the balance sheet is futuristic. On the asset side, it indicates what the firm has available for future use in carrying out its objectives; plating, buffing, etc.

A number of important observations flow from this definition of assets. First, not all assets of a firm are or necessarily could be measured and disclosed in the balance sheet. For example, the human resources of a firm—employees and managers—may be its greatest asset, yet human resources are not reflected in a balance sheet (except indirectly). The reason is simple: they cannot be measured objectively in dollar terms, so they do not enter in the formal accounting process. Consequently, management must effectively and efficiently manage assets which do not appear in the balance sheet, as well as those which do. It is fallacious to assume, therefore, as some managers do, that a balance sheet provides *all* the information regarding *all* the assets of a firm with which management needs to be concerned.

Second, since assets are one of the critical factors utilized in generating income (profits), it follows that there must be some optimum total that any given firm should have, as well as some optimum breakdown of that total into the various types of assets: cash, accounts receivable, property, plant, equipment, etc. These statements are true, although determining those optimum amounts is a difficult process. Nevertheless, one of the most critical management functions is to determine total assets needed, plus amounts needed of individual assets making up that total, which can be used to maximize the income of the firm. Effective and efficient asset acquisition and utilization are critical if successful operations are to result; perhaps they are the most critical functions performed by management.

Third, since assets are at the base of income generation, it follows that the more rapidly they are utilized, holding other things constant, the more profit will be generated. This is true for any asset specifically; it is true for assets in total. Cash lying in a cash drawer does not earn income. A machine sitting idle does not generate income. If profit is to be maximized, therefore, assets must be turned over, utilized in operations, as rapidly as possible. This is a critical truism as far as management is concerned; it represents an underlying principle which constantly should be in the forefront of management thinking. "Keep the assets, individually and collectively, working." Perhaps it should be pointed out that this includes assets not on the balance sheet, as well as those which are.

Fourth, some managers still feel the balance sheet reflects "value," and that they can look at the balance sheet data and see what the "true" value of the firm is. Nothing is farther from the truth. Assets are measured at historical cost (what the firm gave up to acquire them). Acquisition may have been five or ten years ago; prices may have changed significantly in the intervening period. Nevertheless, the assets are stated in the balance sheet at prices prevailing when they were acquired (less the accumulated depreciation total, if any). The reason accountants don't attempt to reflect "value" is simple. Cost is objectively derived; it is not the result of subjective opinion, as is often the case with current values. For most purposes served by the balance sheet, historical cost figures are adequate. For certain other purposes, such as pricing (see Chapters VI and VII), current value (or even anticipated future values) should be estimated and used when possible. In any case, the balance sheet does not reflect current market values, nor does it purport to. It is critical to keep this in mind when interpreting balance sheet data.

Asset Categories

Although the "thread of similarity" running through all assets —future service utility—is critical to see and understand, financial statement analysis and interpretation depends also on understanding significant differences between various types of assets and asset categories. The most important and useful distinction is between current and non-current assets.

A current asset is one which will be utilized, turned over, used up, or consumed during the normal operating cycle of the business, or one year, whichever is longer (from the date of the balance sheet). The normal operating cycle of a business is the length of time it takes to acquire materials, to convert the materials into a finished product or service, to sell the output to a customer, and to collect the amount owing from him. Obviously, the metal finishing industry has more than one operating cycle during the year, so one year can be used as the standard for gauging whether something is a current or a non-current asset.

A number of observations are important regarding this definition of current assets. In the first place, something does not have to be immediately convertible into cash without chance of substantial loss in order to be a current asset. In other words, a high degree of liquidity is not an essential prerequisite for something being a current asset. Work in process inventory frequently is highly illiquid, i.e., it could only be converted into immediate cash at substantial loss. Nevertheless, it is properly classified as a current asset because of the *expectation* that it will be completed in the normal course of business, transferred to the customer, and payment received from him within one year.

In the second place, whether or not an asset has physical existence has nothing to do with its classification as current. Inventory has physical existence; cash, accounts receivable, and prepayments do not. Yet all are current assets.

Non-current assets are those which require a period of time longer than a year in order to be "used up" in total in business operations. The major classification under noncurrent assets is the category of "plant, property, and equipment." These items are the long-lived assets utilized in carrying on the operating activities of the business. Their service utility to the firm, though longer than that for current assets, is limited (except for land), and this limited duration must be reflected in depreciation accounting, where appropriate.

Thus, all assets, current and non-current (except land) are used up in business operations. Some non-current assets frequently are highly liquid (easily convertible to cash), such as land. Some are not, such as highly specialized machinery having no alternative uses. Some current assets are highly liquid, such as cash. Some are not, such as work in process inventories. Assets having physical existence can be found in both categories. Assets not having physical existence can be found in both categories. None of these distinctions is critical, therefore, since none is mutually exclusive. The only meaningful distinction between current and non-current assets that has significance from a management decision-making viewpoint is the length of time it takes to use up the service utility inherent in the asset.

Liabilities and Stockholders' Equity

As there is a "thread of similarity" running through all the assets of a firm, so there is with liabilities and stockholders' equity. All assets must be obtained somewhere, from someone or some organization. The liabilities and stockholders' equity accounts indicate the sources of the assets—where the firm got them. When cash is borrowed from a bank, assets increase as do liabilities. When common stock is issued, cash goes up as does stockholders' equity. Both the bank and the buyers of the stock made resources (assets) available to the entity. In return, the bank must receive the contractual interest on the face of the note, plus a return of the principal amount borrowed, or the firm will be in financial difficulty. Stockholders (owners), as we have seen, must earn a rate of return commensurate with the risks assumed in the undertaking, or they will transfer their capital into more profitable ventures. We discuss the proper relationship of the various components of liabilities and stockholders' equity in Chapter V when we discuss financial statement analysis. Sufficient here to indicate that all the suppliers of assets to the firm do so in anticipation of receiving a return on their investment, and their anticipated returns vary because of the different risks assumed. But all make assets available to a firm to use in its operations, and this is what makes them essentially similar.

There are significant differences between liabilities and stockholders' equity, on the other hand, and even significant differences within the liability category itself. These differences must be understood if the message conveyed by the balance sheet is to be clear and useable to management. Some or all of the following are characteristic of liabilities:

1. Usually there is a sum certain involved. This simply means there is a definite amount which must be paid to the creditor. If materials are acquired on account for $10,000, and assuming no discounts are applicable, then $10,000 must be paid to the creditor; nothing more, nothing less.
2. Usually there is a specified time in the future when the obligation must be paid—a bank note might be payable 60 days from date of note; perhaps a supplier must be paid within 30 days after materials are received.
3. Frequently there is a fixed cost of capital demanded by the creditor. If a bank note carries eight percent interest, for exam-

ple, this is the cost of borrowing to the firm, and it must be paid at the date specified in addition to the principal; nothing more, nothing less.

Because these characteristics attach to liabilities, certain risks are apparent. If the principal amount or interest is not paid at the date specified, the firm might be driven into an illiquid position at best; at worst, it may become insolvent and ultimately face bankruptcy.

None of these characteristics of liabilities is true regarding stockholders (owners). First, there is no sum certain. If an owner wishes to relieve himself of his interest in the firm, he must either find a willing buyer or liquidate the business. In either case, he gets what he can get; i.e., there is no fixed amount involved. Second, there is no fixed point in time when he gets his capital investment returned. He receives it when he disposes of his interest in the firm, either by finding a willing buyer or by liquidation. Third, and finally, his rate of return is determined by the profits generated by the firm. If the firm is successful, his return may be high. If the firm is unsuccessful, his loss may be large. The owner is the primary risk taker. He is a residual claimant. He bears the major and primary burden when income is low; he reaps the rewards when income is high.

The major distinction between current and long-term liabilities is consistent, and must be, with the distinction between current and non-current assets. A current liability is one which must be liquidated, paid off, within one year of the balance sheet date, usually by use of a current asset. A long-term liability is one whose payment date is longer than one year removed from the balance sheet date.

Income Statement

An income statement reflects the results of the operating activity of the firm for a given period of time. The period can be of any length: a week, a month, a quarter, a year. An income statement is like a moving picture which shows the effects of the operating transactions and events of the firm for the period in question. Operating events and transactions are those related to the economic activities of the firm, i.e., the activities associated with providing the product or the service to the customer. Following is a representative income statement, reflecting only the major categories. Many income statements would show in detail the individual accounts under production expense, selling expense, general and administrative expense, and taxes expense. For our purposes here and those of Chapter IV, the points are more clearly and understandably presented using major categories only.

METAL FINISHING FIRM
INCOME STATEMENT
For the Year (week, month, quarter)
Ended December 31, 19—

Sales Revenue		$709,607
Less: Production expenses (cost of goods sold expense)		449,181
Gross profit margin		$260,426
Deduct: Selling expenses	$ 24,296	
General and administrative expenses	156,055	180,351
Income before income taxes		$ 80,075
State and federal income taxes		21,075
Income before extraordinary items		$ 59,000
Extraordinary items:		
Fire loss (net of tax savings of $10,000)		30,000
Net income		$ 29,000

Since we have taken precautions throughout this monograph to protect the identity of specific firms, the income statement data above were not taken from the records of any firm. However, the relationships among the items in the statement, and those items relative to the sales revenue figure, approximate the relationships existing for some firms in the industry. This adds realism to the remaining comments of this chapter and those in Chapter V.

Sales Revenue

There are a number of ways to define sales revenue, all of them accurate and all of them useful for certain purposes. Sometimes revenue is defined as the total amount collectible from customers for the delivery of goods and services during the period covered by the income statement. Some define it as the monetary measure of a service rendered. To us the most useful definition of revenue is the *measurement* of the change in gross assets brought about by the transfer of the goods or services of the firm to its customers during the period of time covered by the income statement. Revenue has no existence, as does cash or accounts receivable or a machine. Revenue simply is a *measurement,* nothing more or less, but what it measures is critical for successful operation, i.e., the gross asset increase resulting when goods or services are transferred to customers. When finishing services are transferred to a customer, say for $15,000 on account, accounts receivable (a current asset) increases $15,000; revenue measures that increase. The nature of the asset received by the firm in exchange for the goods transferred to the customer, or the services performed for him, is irrelevant. Cash might be received immediately; a promise to pay in the future might be received (accounts receivable); even a machine or a piece of land might be received (though this would be unusual). It makes no difference; revenue would result in all these instances.

Expenses

Another important classification in the income statement is expenses. As is true for revenue, expense is defined in a number of ways. Expense can be defined as all costs incurred intentionally to produce the revenues of the period, or as one of the amounts (negative factors) deducted on an income statement in arriving at net income. The best definition of expense, we feel, is the *measure* of the cost of assets used up in producing revenue. Expenses are good, therefore. Rarely do we get something for nothing. Expenses *measure* service utility consumed and gone from the firm forever, *which was productive of revenue.* Expenses are measurements of asset decreases efficiently and effectively utilized in generating revenue. Of course, such decreases take on a large number of specific forms: raw materials, labor, depreciation, administrative salaries, and many others. The critical point is that they are related to, or causal of, the revenue of the period. Therefore, they must be associated with that revenue (deducted from it) in order to fairly determine net income for the period covered by the income statement. If the service utility remains in the firm, it is an asset. If it doesn't, it is either an expense or a loss.

Losses

Once the concept of expense is understood, losses are easy to comprehend, although admittedly difficult to measure at times. A loss *measures* service utility consumed (asset decreases) which is not causal of revenue. Losses are bad. They are not productive. They should be held to the minimum possible. For example, using more raw materials, more labor, more overhead, or more of anything else than reasonable standards dictate ought to be used in the circumstances, result in losses and are to be avoided. Such avoidance is aided by the establishment and utilization of reasonable cost standards against which actual results can be compared. And, of course, losses may result from more irregular and unusual circumstances such as fires, condemnation, and the like. In any case, this discussion drives home a point made earlier; namely, that two of the most critical functions of management, which must be performed well if operations are to be successful, are the *acquisition* and *utilization* of assets.

Net Income

Net income also is defined in many ways: as the resulting figure when expenses and losses are deducted from revenue; as the earnings which management has produced during the period. Our preferred definition is that "net income" *measures* the change in net assets brought about during the period covered by the income statement as a result of the operating activities of the firm (net assets equals assets minus liabilities). Assets can change for a variety of reasons. Net income is not net assets; net income *measures* certain net asset changes—those brought about by providing a product or a service for the customer, such as finishing services. Net income also measures the increase in stockholders' equity during a period of time from operating activities, reflected as an increase in retained earnings.

A Few Warnings and Conclusions

Note throughout this discussion that we have not mentioned the inflow of cash. Some managers equate revenue with the immediate inflow of cash, and expenses and losses with the immediate outflow of cash. Consequently, they feel that net income should measure the net change in cash that took place during the period covered by the income statement. All such feelings are fallacious and to be avoided. Ultimately there will be cash consequences regarding all of the classifications in the income statement, to be sure, but not necessarily during the period covered by the income statement. Revenue is recognized in the income statement in the period when the goods or services are transferred to the customer. Cash may be received before, during, or after that period. Expenses are recognized in the period when they are productive of revenue. Cash may be paid out before, during, or after that period. For example, a machine may be purchased in 1976 for $10,000 cash, but that entire amount would not properly be considered as an expense in 1976, if we assume a useful life of ten years for the machine. The $10,000 would be allocated over the ten-year period in some appropriate fashion as depreciation expense. (The next major section of this chapter covers the "funds statement," which will help show the relationship of income statement classifications to cash changes). What we have been describing here is the accrual basis of accounting, and it is the proper method of reporting data in the balance sheet and income statement.

Most businessmen will state that their primary objective is to maximize income or projects and, in fact, we state this in the chapters on pricing as the basic objective of the firm. Nothing is wrong with this at all, provided we understand clearly what it means. Long ago, a man said that "the primary objective of businessmen is to make two blades of grass grow where one grew before." What this means, simply, is that the real goal of businessmen is to utilize assets in such a way that they end up with more assets than they started with. The emphasis is on "assets," as our emphasis in the definitions above, was on "assets." Thus, the basic objective of the finishing firm manager ought to be to maximize the growth in net assets from operating activities. This is what he should keep in mind when he thinks of maximization of income or profits. This will keep his attention focused in the right direction, i.e., on the efficient and effective acquisition and utilization of assets.

The Funds Statement

Many of the firms in our survey made use of the balance sheet and income statement, although some did not make very effective use of them. Few used the funds statement at all. Yet, the funds statement must be presented in the published reports of publicly held corporations, and it is now viewed in many quarters as equal in importance to the income statement and balance sheet. It must have merits, therefore, and deserves consideration here.

What Is a Funds Statement?

The balance sheet displays the financial position of a firm *at* a given point in time. Looking at it, we can see the balances in various accounts at that time: the cash in bank, accounts receivable balance, book value of machinery, and amounts owing to various creditors, among others. If comparative balance sheets are presented, somewhat more information is available. By simple subtraction we can ascertain by how much the various items in the balance sheet changed during the period. For example, if the cash balance was $12,000 at the end of last year, and $18,000 at the end of the current year, then we know cash increased by a *net* amount of $6,000. What we cannot see, looking at the balance sheet alone, is what activities and events brought about this *net* change of $6,000 in cash. In a word, then, the funds statement's purpose is to reflect by major category what brought about the change in funds during the period covered by the statement.

Definition of Funds

"Funds" can be defined in a number of ways: as cash, as working capital (current assets minus current liabilities), or as all financial resources. The first problem then, is to select the concept of funds to be utilized in preparing the funds statement. Though each definition has merit for certain purposes, for management decision-making purposes in finishing firms,the best concept of funds to follow is cash, for a number of reasons. Maintaining and utilizing cash effectively is one of the most difficult and complex tasks of management. Second, most events and transactions ultimately pass through or have an effect on the cash account. Third, a number of erroneous feelings held by some managers (such as changes in cash equal net income for the period) are cleared up when funds are defined as cash. Fourth, more events and transactions affect cash during a period, generally, than any other account.

When cash is selected as the definition of funds, the purpose of the funds statement then is to reflect, by major category, those events and transactions occurring during the period which brought about the *net* change in the cash balance. It is a "where got, where gone" statement of cash changes: where we got the cash during the period, and what we used the cash for.

Preparing a funds statement is a fairly complex matter requiring the help of experts in the area, so we will spend little time with the techniques of preparation. A brief discussion is important for managers to understand the types of information provided and questions answered by the funds statement, i.e., how the statement can be used most effectively.

Following is a representative sample of a funds statement in a format we believe is of greatest usefulness to managers of finishing firms. We have utilized the data from the previously presented balance sheet and income statement and additional assumed data where necessary, in order to show the interrelationships of the statements.

Informational Content of the Funds Statement

First, the funds statement makes it clear why income before extraordinary items ($59,000) did not increase cash by the same amount (cash only increased $4,000). Second, it makes clear the relationship between the income figure of $59,000 on an accrual basis and cash flow from operations of $75,400. Some items affect income determination but have no effect on cash during the current period. Such items are, therefore, properly utilized as adjustments to income to derive cash provided by operations. Third, this format indicates the cash effects of major categories of events and transactions, such as financing activities. In other words, it shows the major sources of cash and the major uses of cash during the period. Fourth, and finally, it displays important subtotals, such as cash available for expansion.

We will consider the detailed analysis of the funds statement in Chapter V. Here we will indicate the major informational content of the funds statement on a cash basis.

METAL FINISHING FIRM
FUNDS STATEMENT - CASH BASIS
For the Year (week, month, quarter)
Ended December 31, 19—

Income Before Extraordinary Items			$59,000
Add: Adjustments necessary to convert income from an accrual to a cash basis:			
Increase in wages payable		$500	
Increase in accounts payable		600	
Decrease in prepaid expenses		400	
		$1,500	
Expenses per the income statement not affecting cash this year:			
Amortization of goodwill	$200		
Depreciation expense	17,000	17,200	18,700
			$77,700
Deduct: Adjustments necessary to convert income from an accrual to a cash basis:			
Decrease in income taxes payable		$300	
Decrease in social security taxes payable		100	
Increase in inventories		900	
Increase in receivables (net)		1,000	2,300
Cash Flow from Operations			$75,400

Extraordinary Items			
Cash received from the insurance company for fire loss (including tax saving of $10,000)			15,000
			$90,400
Financing Activities			
Cash derived from issuance of mortgage bonds payable	$30,000		
Less: cash used to pay a three-year bank note	6,000		
Net cash in-flow from financing activities			24,000
			$114,400
Cash dividend paid during the year			5,000
Cash Available For Expansion			$109,400
Expansion Activities			
Cash used to expand plant, property, and equipment:			
To acquire machinery and equipment		$30,000	
To acquire land		12,000	
To acquire buildings and make major improvements and renovations		63,400	
Total Cash Used For Plant Expansion			105,400
Increase in Cash Balance During the Period			$ 4,000

Beginning of the period cash balance	$3,000
Increase in cash balance during the period	4,000
End of the period cash balance	$7,000

1. The $75,400 cash provided by operations indicates that revenues increasing cash exceeded the expenses utilizing cash by that amount during the period. When assessed over several years, this most important source of cash indicates the extent to which operating activities have generated more cash than they have used. This excess can be used to pay dividends, repay long-term debt, and acquire buildings and equipment. If this figure is small or negative, it indicates financial trouble for the firm.
2. Cash derived from long-term debt or common stock indicates how the firm is making use of long-term sources of financing for a firm, as compared with the short-term nature of cash from operations.
3. The acquisition of plant, property, and equipment represents an important use of cash. Assets must be replaced as they wear out, and additional assets must be acquired if the firm is to grow.

Overall then, the funds statement (particularly the comparative funds statement) discloses changes in the liquidity of the firm and changes in the structure of a firm's assets, liabilities, and stockholders' equity. Such changes, and the reasons for them, must be observed and understood if management is to comprehend fully how the firm got to its present financial position, and to predict how its position is likely to change in the future.

Although the balance sheet, income statement, and funds statement provide useful information as to the present status of a firm and how it got there, these statements along with other data are also useful for planning purposes, i.e., for looking into the uncertain future, establishing objectives, and taking actions to reach these objectives—all necessary to successful operations.

Planning — A Critical Factor for Successful Operations

While it is true that planning techniques usually are more highly developed in large well-run organizations than they are in similar businesses of lesser size, these practices are equally critical to medium- and small-sized firms. There may, in fact, be more need for smaller companies to emphasize these tools. They are under greater temptation to try to conduct operations on a day-to-day basis, and their personnel and organization structure may not be adequate to allow decisions to be made otherwise than from day to day, as problems arise.

Managers of small businesses may feel that they are so closely associated with all aspects of their operations that there is no need for planning, or that efforts in this direction are impractical. These attitudes are reflected by owners and managers of many of the firms in the metal finishing industry. Only 48 percent of the respondents indicated their firms used budgeting techniques at all. Of that 48 percent, 56 percent indicated budgeting was used as an aid in price setting, and 83 percent indicated budgeting techniques were used for cost control purposes.

If the industry is to enjoy continued growth, and if individual firms in the industry are to prosper, these attitudes need to give way to new ones in which managers consciously consider and establish profit goals for their companies. They should develop practical workable budgets and direct their efforts and those of their employees toward the accomplishment of sales, cost, and profit goals. They should see that meaningful measurements are continually made to determine the extent of and reasons for deviations from the charted paths toward these goals, and they should then take the necessary actions to assure reaching the established targets. In a very brief form, this is the nature of the management process. It applies to all companies without regard to size. It applies to service organizations and to job shops. Although budgeting is critical to both pricing and cost control, we cover it briefly here only as a basis for pricing as discussed in Chapters VI and VII. Cost control is critical, of course, but, as indicated previously, it constitutes a large area by itself and will not be covered in this monograph. However, the basic discussion of budgeting in this chapter is as applicable to cost control as it is to pricing decisions, should some readers decide to pursue it further from the cost control angle.

The planning process should be considered in terms of both long- and short-term spans. The owners and managers of metal finishing firms must decide on many matters such as plant size, plant location, types of metal finishing service to provide, and industries to serve, all of which have long-run ramifications. Plans in these areas are most strategic and must be made with great care. Within the setting conditioned by decisions of these kinds, managers should develop short-run financial plans — annual budgets.

The annual budget should detail the production and sales goals, with the cost, revenue, and profit consequences that are to be used to direct and control business activities for the impending year. It should reflect the financial objectives and the plans that have been developed for reaching these objectives.

Budgets - Their Benefits

The process of preparing a detailed plan or budget for the coming year is, in itself, a rewarding activity. The process forces the manager or several of the management personnel to go through a "dry run" of likely developments. In this way many of the critical problems to be faced later are brought to light, and consideration and study of the appropriate action may get underway well before the actual problems arise and must be solved. Possible courses of action may be developed and appraised, and sound decisions may be made by recognizing many factors which might have been overlooked if the problem were allowed to develop unexpectedly.

Preparation of the budget should involve individuals who have responsibility for sales, for production, for personnel, for plant maintenance, and for all the other functions to be performed. Plans for each of these activities are evaluated in dollar terms and meshed to form a total plan which will lead to a suitable result in terms of profit and financial position. Each person participating will more clearly see his role in contributing to these results, and this awareness should further motivate him to fulfill his part of the total plan.

The use of an annual budget assists in coordinating efforts toward the central focus of profit maximization. In terms of pricing policies, it encourages teamwork among those responsible for such decisions; it prevents needless duplication of effort and ensures that people will not be working at cross purposes. Each person's area of responsibility becomes an integral part of the overall objectives of the firm, rather than an end in itself.

Components of the Annual Budget

The most critical element in the annual budget is the plan for sales. It is the foundation for all of the subbudgets that combine to make the total plan. The sales budget should reflect the dollar volume of business anticipated for each finishing service and the total revenue target for the firm. In some cases, the volume may be projected in terms of customer groups. The degree of detail should depend upon the variety of types of finishing involved, the variety of methods used, and the variety of customers served. The budget should be detailed enough to facilitate reasonable accuracy and to permit goals to be set in terms which will be useful in measuring success and taking corrective action.

Sales budgeting is a difficult task in any business, and perhaps it is even more difficult in job shops serving other industries rather than the public directly. Although the temptation is strong to refuse to engage in sales budgeting because no one can foresee what may happen and a feeling of control over critical factors is felt to be limited, job shop metal finishers cannot afford to be without sales budgets. The more competitive the industry is and the more difficult the job of budgeting sales, the greater is the need for sales planning. It is critical if the target rate of return is to be achieved. It is critical to attain the most profitable mix of finishing jobs. It is critical in order to have something to compare with actual results in order to observe problem areas. Then corrective action can be taken, where warranted and possible.

The top sales executive should be primarily responsible for preparing the sales budget, his work being subject to review and consideration by the other members of the top management group. In preparing the budget he will want to consider past sales records, general economic and competitive conditions, changes in prices, target rates of return, estimates by other sales people, and market research studies of the industries or firms to be served. Close personal contact with purchasing managers of potential customers is helpful. All of these kinds of information should be considered and the results analyzed and adjusted to the extent possible and necessary to accomplish the basic profit objective of the firm.

Once a sales budget has been completed, it then serves as a guide for preparing the other major components of the budget. The most important of these is the production budget. The production budget indicates the number and types of operations to be performed. These data must then be converted to show future requirements for material, personnel, plant capability, and service. These requirements are, in turn, measured in dollar terms to determine fund requirements and total planned costs of production. The figures used here should be based on an assumption that reasonable efficiency in operations will be attained and that the prices used will be those prevailing during the period for which the budgets are being prepared.

Budgets for selling and administrative expense should then be prepared. We will see in Chapter VII how these are utilized for pricing decisions. When these things are done, the first approach to a complete annual operating plan has been made. This operating plan should be summarized in the form of a projected income statement for the coming year, and a balance sheet projected as of the end of the coming year.

These "first approach" statements should be examined closely to see the extent to which their fulfillment would provide a suitable financial result. Is the profit figure acceptable? Is it a reasonable return on the capital invested? Is the resulting balance sheet acceptable? If any of these questions cannot be answered in the affirmative, the budget process is not complete. Each factor in the plan should be examined to discover what revisions are possible to meet the firm's objectives. Revisions may be minor, or it may be necessary to repeat the entire process of planning.

A good accounting system is a prerequisite to good budgeting. The account structure used for general accounting purposes provides at least the starting point for the budget process. The Uniform System of Accounts prepared for the National Association of Metal Finishers can be used in this regard.

"Hopes," Not "Expectations," Critical

Budgets should reflect the *reasonable hopes* of management for a period, not what they expect will *actually occur*. Budgets should reflect action plans, what management intends to do to make things happen as they want. If, as is sometimes the case, budgets reflect expectations rather than *reasonable desires,* a number of advantages are lost. Under such an approach, management is not likely to maximize long-range profits. It is more useful, for example, to know why profits were or were not what they ought to have been than to know whether or not they were equal to expectations. When budgets are based on reasonable desires, the analyses and interpretations of deviations can indicate which ones are controllable and which are not. Management can then concentrate on the controllable deviations to ensure that the firm attains its objectives, or at least comes as close as possible. Management also can analyze and interpret the uncontrollable conditions as they arise and adjust plans and procedures accordingly.

Such an approach can also help gauge whether or not the objectives are reasonable. If the goals are never closely approached, it might be concluded that hopes are unrealistic or that the desires (demand) of the market are different from what planners anticipated. In this event, an adjustment of product mix (relative weights given to each type of finishing) may be necessary in order to attain the overall return on investment or other objectives that management has set up as desirable. Either a temporary or permanent change in the market would require some alteration in plans.

The Accounting System

In order to provide the type of data needed for the calculations and reports we have discussed and will discuss in the following chapters, a relatively complete and accurate accounting system is needed. Certainly, as indicated, the Uniform System of Accounts prepared for the National Association of Metal Finishers is a good starting point. More will be needed, however, if revenues, variable costs, and traceable fixed costs are to be broken down by finishing line for budgeting and pricing decision purposes.

Modern technology promises to make possible a quantum jump in plating information recording and accounting. Now available to the industry is electronic data entry from the finishing shop directly into a minicomputer. The time a finisher spends on individual job operations, revenues, variable costs by finishing line, and other data needed can be entered through a terminal. The need to manually record such information on paper is eliminated. Besides reducing messy paperwork, the electronic process is faster and much simpler. Manual clerical tabulations of job costs are also avoided.

Such systems have been implemented in related industries for several years, and at surprisingly low costs. The key element is the low cost minicomputer, which several progressive finishing companies are now acquiring. Managers can be provided with information such as production costs and labor efficiency reports on a weekly, daily, or hourly basis if they wish. Because much of the clerical work otherwise required is eliminated, these systems hold great hope for the manager who wants to progress, but dislikes the attendant growth in paperwork and manual bookkeeping. Moreover, recording job time data is only one of a myriad of administrative details that can be computerized. Job expense information for materials, maintenance, rework, etc., all can be entered electronically and accounted for by the computer and then reported on an up-to-the-minute basis to the manager. Employment records, such as vacation, sick leave, and payroll deductions, can be computed easily and quickly from such basic data as time reports. Certainly management should look into the possibility of computerizing the most important aspects of their accounting system, even if placing the entire system on computers is not feasible.

Whether or not electronic data processing equipment is used, many firms still need to upgrade their accounting systems. Our observations indicate that significant numbers of firms, at best, have only the most rudimentary of systems. Often these systems do not provide adequate information for planning and control purposes; frequently, little if any analysis and interpretations are provided to the managers. Most of what we recommend in this monograph can only be accomplished if sufficient relevant data are provided by the accounting system. Those firms too small to afford the necessary personnel and equipment to do this will have to find some other way, such as sharing an accountant employee, or engaging one or a few practicing accountants in the area to do the necessary work. By handling the accounting functions of a number of finishing firms, such an accountant can gain expertise in providing the types of data needed for successful operations. Also, it may be possible for a number of firms to share computer facilities.

IV

COSTS — THEIR NATURE AND CLASSIFICATION FOR DECISION MAKING

No matter what the nature of the decision facing the businessman, a basic understanding of the nature of costs and how they behave under differing circumstances is essential to informed decision making. Pricing decisions which must continually be made, vigorous competition, and the increasing rigidity of costs associated with mechanized production make essential an understanding of how costs and volume are related to profit. It is also important to understand the nature of costs and how they react under varying circumstances to comprehend fully the discussions in the following chapters. These discussions lead to many of the ultimate conclusions reached in this study.

Finishing costs may be classified in many ways: direct vs. indirect, historical vs. current, controllable vs. noncontrollable, variable vs. fixed, to name a few. Although each classification is relevant for certain types of decisions, the most important distinction among costs for most business decisions involves the classification in terms of their relationship to volume of production or sales, i.e., the way costs behave as volume changes. Certainly this is true regarding cost relationships in the finishing industry, so we will concentrate on the variable/fixed classification.

Incremental cost and revenue analysis is the essence of choosing between alternatives. "Incremental" simply means the item will change between alternative courses of action. If something will change, it must be considered in making a decision between alternative courses of action; if not, it may be ignored. For most decisions, incremental costs will equal variable costs, although there are exceptions of course. For our purposes here, however, we will use the terms interchangeably.

Variable Costs

Many costs are directly related to production or sales volume and vary in total dollar amount with changes in volume on an approximately proportionate basis. These are variable costs. When volume of production increases 10 percent, variable costs will increase 10 percent, and so on. This means that costs which are truly variable will remain constant per unit, and for most decisions made by the finishing manager, they should be considered in this per unit sense. Examples are raw materials, direct labor, sales commissions, delivery costs, and electricity used in production. Figure 5 reflects the way variable costs react to volume changes. The diagram shows that the cost to produce 1,000 units is $100, or $.10 per unit. To produce 2,000 units costs $200, or $.10 per unit, etc. Such a cost would be truly variable since the total dollar amount varies in exact proportion to volume changes while remaining constant at $.10 per unit.

Fig. 5. Variable Cost Behavior as Volume Varies

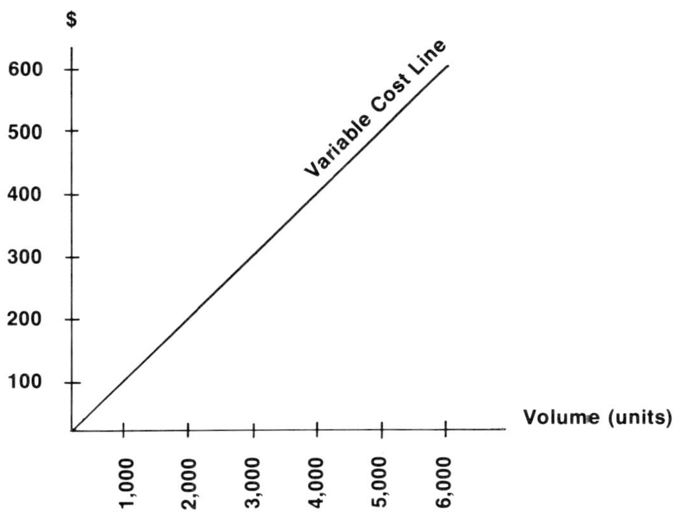

Material costs usually vary proportionately with volume, if the mix of jobs and the length of production runs remain relatively constant, although discounts may increase as larger quantities of materials are purchased when volume increases. As a consequence, when volume increases 10 percent, material costs may increase somewhat less than 10 percent. Nevertheless, the variable cost grouping is justified for our purposes here.

Other variable costs have a tendency to creep up per unit as production decreases, though they drop in total amount as volume declines. Direct labor costs, for example, are basically variable in nature, but restrictions in labor contracts tend to cause efficiency to lessen as production decreases, resulting in an increase in direct labor cost per unit. Payments which continue for a time after the worker is laid off, such as Supplemental Unemployment Compensation, also tend to move direct labor somewhat toward the fixed cost category. In addition, even though one ideally can expect additional personnel to be hired in response to a larger volume of activity, and that payroll will decrease as volume of operations is reduced, management policy may make some basic amount of labor costs essentially fixed. For example, some basic portion of the labor force might be maintained intact regardless of volume fluctuations, in order to avoid other costs such as those related to dismissal, rehiring, retraining, and unfavorable unemployment insurance rates. The costs associated with the basic labor force must be maintained at all times, therefore, and a basic fixed portion of labor costs may be expected to exist. Nevertheless, management, in response to changing volume, should be able to keep those costs relatively constant per unit, and thus the variable cost grouping is justified.

Fixed Costs

Fixed costs are those which stay relatively constant in total dollar amount over given relevant ranges of production activity. Thus, they would be the same at the bottom of the relevant range as they are at the top, or at any volume level between the two. They are the costs of getting ready to do business, including the costs of promoting and influencing future business. For example, when a business organizes and acquires buildings, machinery, and equipment, decisions are made on the basis of the volume of activity which the firm anticipates. Although the amount of the investment will depend on the anticipated volume of business, the initial outlay is necessary before the firm is ready to do business; and the plant and equipment must be maintained regardless of the volume of business at any particular time. Such costs relate more to the passage of time than to the volume of activity within a specified period of time. Once these initial outlays have been made, the investment is "sunk" and the attendant depreciation charges are uncontrollable. The investment cannot be retrieved under usual circumstances without substantial loss. Moreover, the acquisition of such items creates the requirement of continuing cash outlays for taxes and insurance. These costs frequently continue even if the plant is completely shut down. They can be reduced only over an extended period as capacity is gradually reduced by retirements, and acquisitions and replacements are restricted. Thus, a significant portion of fixed costs results from the initial provision of the capacity necessary to operate. Figure 6 reflects the way fixed costs react to volume changes.

Fig. 6. Fixed Cost Behavior as Volume Varies

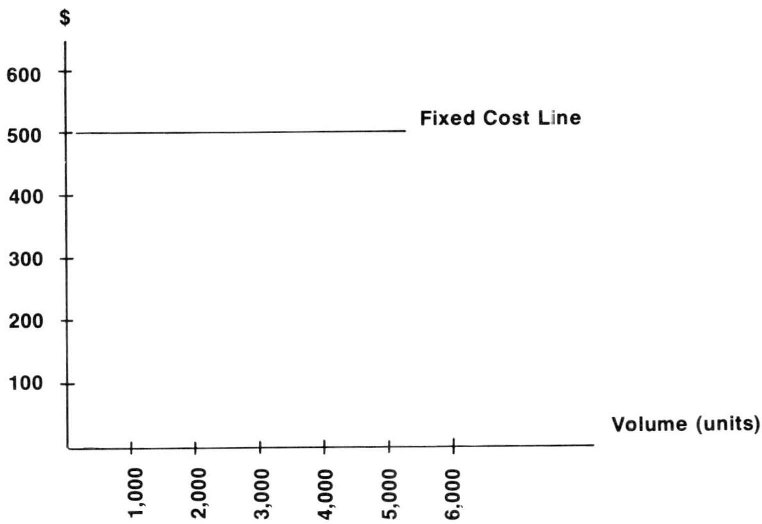

Costs in Figure 6 are truly fixed, since the total amount remains constant at $500 regardless of whether volume is zero, 1,000, 3,000, 5,000, etc., units produced. A critical point in this regard should not be overlooked. With 1,000 units produced, it costs 50 cents per unit; at a production level of 5,000 units, per unit cost drops to $.10, etc. Thus, as volume increases, fixed costs per unit decrease since total fixed costs remain constant. This emphasizes the importance of operating at a high activity level, particularly in those businesses having a high proportion of fixed costs to total costs. This certainly is true for a number of types of metal finishing operations, particularly where mechanized equipment is used heavily.

Most of the conclusions reached above also apply to the outlays for salaries of operating personnel, although these costs are more controllable than those arising from the acquisition of plant facilities and equipment. A minimum number of operating and office employees must be retained regardless of the volume of output. To this extent these costs are fixed, although only within a known range of volume. As volume increases, a point is reached where additional personnel are necessary. Thus, as volume fluctuates, salaried personnel costs fluctuate, but in sizeable discrete steps rather than in small continuing increments.

These costs may be treated as fixed costs, although it is recognized that there are levels of volume beyond which they will change. This kind of cost behavior is often referred to as a semi-fixed pattern.

Expenditures for such things as advertising, personnel training, safety programs, and special research projects do not vary with volume, except in the very long run. Although these costs are important and justified, they can be reduced drastically for coming periods of time if such reduction is necessary for survival. For any given period, however, once agreements have been reached and contracts signed, such costs usually cannot be reduced. Costs of this type, therefore, should be considered fixed. In some instances, however, it may be useful to subclassify them as programmed or managed costs to emphasize their discretionary nature.

Mixed Costs

Some costs fluctuate in small continuing increments as volume changes. They are mixed or semivariable costs — i.e., they reflect a mixture of both fixed and variable patterns. Special consideration and attention may be needed to determine the behavior of these costs. Costs of water, electricity, and maintenance are likely to change as production levels change, but they will continue at some lower level even if finishing operations are temporarily shut down. For such costs as these, it may be useful to set up both a variable portion and a fixed portion and to deal with them in separate segments. Figure 7 reflects the reaction of a cost with mixed-cost tendencies to volume changes.

Fig. 7. Mixed Cost Behavior as Volume Varies

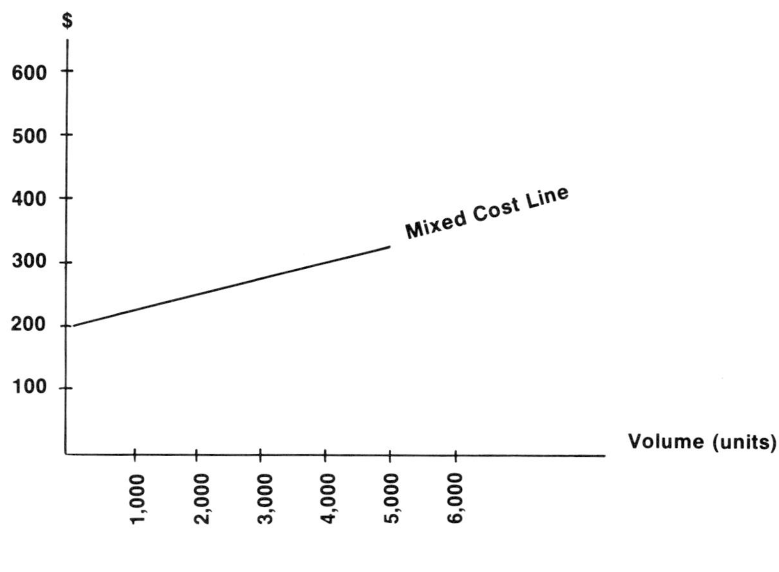

This figure shows that at any given volume level, $200 of the total costs incurred is fixed and the remainder is variable. At zero units produced, $200 of costs would be incurred, all fixed. At 2,000 units produced, $250 of costs would be incurred, $200 fixed and $50 variable (2½ cents per unit). At 5,000 units of production, total costs incurred would be $325, $200 fixed and $125 variable (2½ cents per unit).

How to Determine the Nature of a Cost

Most finishing company managers should be able to classify their costs into fixed and variable categories by simply considering the nature of each cost and observing how it reacts as volume changes. Experience and observation would indicate that direct labor and raw materials are variable costs, as a general rule. Depreciation on the building is likely to be fixed.

In some instances, however, the plotting of costs of various past levels of production activity in relation to volume (scatter diagrams) may be used (see Figure 8).

Fig. 8. Scatter Diagram — Costs at Various Volume Levels

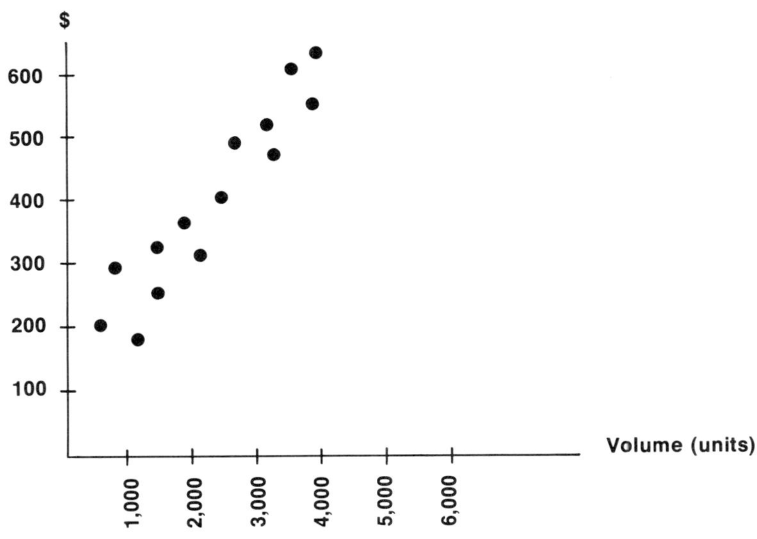

Then a straight line is fitted to these points:

Fig. 9. Interpreting Scatter Diagram Data

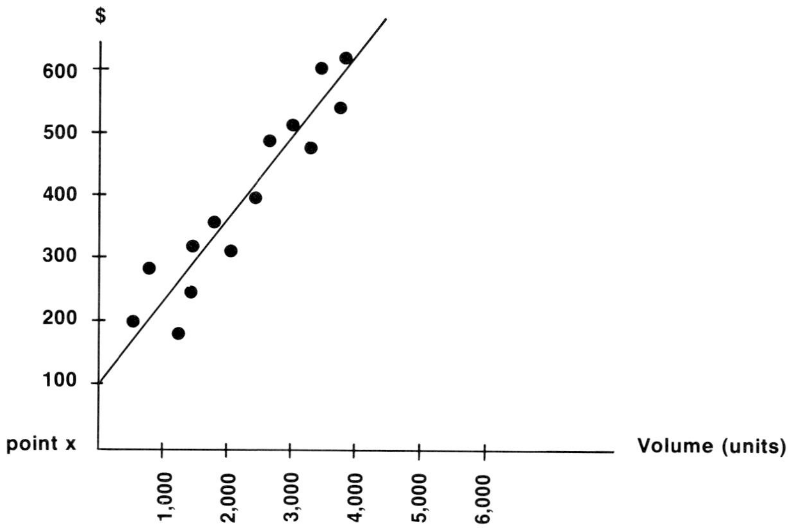

Where the line crosses the "$" axis (in this case $100), it indicates the fixed costs portion of total costs, the remainder being variable at the different levels of activity. Had the line been parallel to the "Volume (units)" axis at the $100 level, then all costs would have been fixed; had the line crossed at point "x," the costs would have all been variable.

Fairly simple mathematical calculations can be utilized to arrive at the same conclusions. For example, assume the accounting records indicate that total costs are $200 at the 1,000 unit volume level, and $410 at the 3,100 volume level. Then, total costs increase $210 ($410 – $200) over a change in volume of 2,100 units (3,100 – 1,000).

$$\frac{\$210}{2{,}100} = 10 \text{ cents per unit variable costs.}$$

Fixed costs must therefore be $100 ($200 – 1,000 × 10 cents) or ($410 – 3,100 × 10 cents).

The foregoing examples give a very brief and limited explanation of the problem of analyzing the behavior of historical costs in finishing operations. More detailed and sophisticated techniques for handling the problem are discussed in many current publications. In any case, through observation and analysis, finishing company managers can separate total costs into their variable and fixed components accurately enough to aid in the types of decisions required for more successful operations.

A Few Warnings

It should be recognized that the patterns of behavior of some recorded costs reflect arbitrary decisions regarding accounting methods. For instance, recording depreciation on a straight-line basis gives the impression that such a cost is fixed (totally unresponsive to volume changes), whereas accounting for depreciation on the basis of machine usage causes the amount to reflect a variable pattern, moving in direct relation to volume. It is the nature of the item, not the way it is calculated, which determines whether it is fixed or variable. Therefore, in accounting for depreciation or other changes where alternative accounting methods are available, the method chosen should be the one which most nearly reflects the expiration of service utility or of the benefit received. If this is done, fixed and variable groupings based on reflected cost patterns will provide reasonably accurate measures for decision making in the finishing industry. If other methods of accounting are better for income tax or other purposes, appropriate adjustments will have to be made.

Of course, all costs are variable in the very long run. Plans must be drawn for short periods (yearly, or less), however, and many decisions should reflect the expectations of the short-run behavior of costs. Fixed and variable classifications provide a practical tool in managing a business, even though they are not completely sound as theoretical concepts and even though a high degree of accuracy, in the mathematical sense, cannot be expected.

As discussed later in this study, the classification of costs provides the basis for important operational analyses. The use of these classifications usually provides no verification of their accuracy. Faulty classifications may lead to poor decisions and serious consequences. It is thus important that careful attention be given to devising the classifications and making sure they remain reasonable.

Cost behavior reflects particular circumstances with regard to the nature of the finishing processes, management policies, accounting practices, and cost incidence. As these factors change, cost behavior patterns may be expected to change. It is, therefore, essential that cost behavior and cost classifications be reviewed periodically, if not continually, to assure a sufficient degree of accuracy in the analyses of finishing operations which use the classifications.

Cost Analyses for Finishing Operations Using Fixed and Variable Cost Classifications

The use of cost behavior information for pricing decisions is considered in Chapters VI and VII. We consider here only the basic profit sensitivity analysis or break-even analysis made possible by behavioral cost classifications. Having examined the relation between costs and volume, we now consider the interrelations of these factors with revenue and profit.

The interrelations may be simplified in a single chart as given in Figure 10.

Fig. 10. Cost and Volume as Interrelated with Revenue and Profit

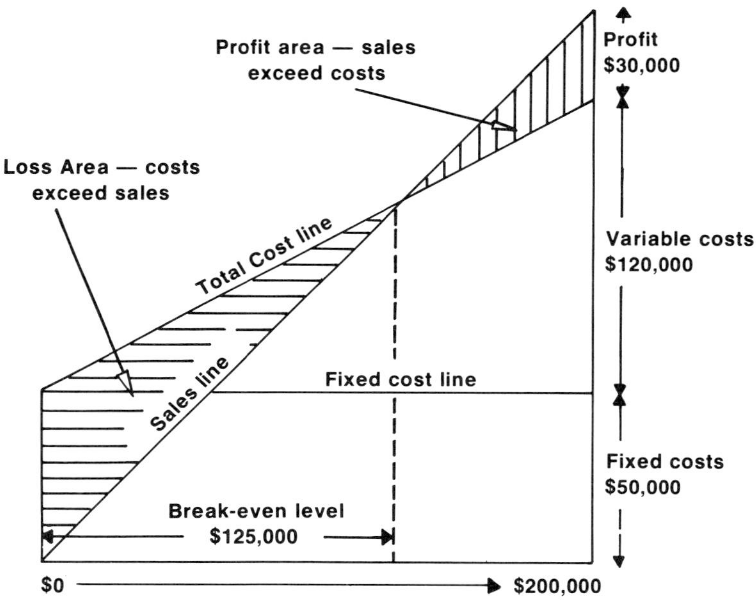

This diagram shows the level of profit or loss that may be expected for a typical finishing company at any sales level up to $200,000. Fixed costs are assumed to be $50,000. This amount would include some items of departmental and factory overhead, most of the building and occupancy costs, and selling and administrative expenses. Variable costs are expected to be 60 percent of sales, or $120,000 if sales reach $200,000. The sales volume necessary to break even is $125,000. This may be calculated by dividing the $50,000 fixed costs by 40 percent, the contribution margin rate (contribution margin divided by sales). At this break-even level, variable costs amount to 60 percent of $125,000, or $75,000, leaving $50,000 to cover exactly the fixed costs. If sales fall short of $125,000, a loss may be expected. A profit should result from sales levels over $125,000. Expected profit (before income taxes) may be calculated readily by comparing the contribution margin rate (40 percent of sales in this case) with the $50,000 fixed costs. Profits, before taxes, may be expected to change by an amount equal to 40 percent of any change in sales above the break-even point.

The break-even analyses already discussed are particularly useful in evaluating overall finishing operations. Several critical assumptions are involved and must be recognized, however. In addition to assuming the accuracy of cost behavior, we assume a constancy of efficiency, unchanging input and output prices, and a stable mixture of finishing jobs. Commonly, these factors are continually changing, so the analyses must be updated frequently and used with discernment to prevent misinterpretations.

Break-even analyses are especially useful when applied at departmental levels or in terms of the particular type of finishing operations. Practical considerations will determine the extent to which the analysis may be applied at segmented levels. We discuss this more fully in Chapter VII.

Example Applications of Variable/Fixed Cost Classifications

An important decision which frequently comes up is whether a finishing company should operate more than one shift per day. Certainly it is not as simple as saying, "If sales justify multi-shifts, do it; if not, don't." Obviously, if sales cannot be obtained for more than one shift, then only one shift will be operated. However, if analyses indicate that additional shifts would be profitable *if* sales could be obtained, this would encourage additional marketing (sales) effort. More importantly, proper analysis would indicate the direction the sales effort should take, i.e., the type of customer and/or finishing operation which would be the most profitable and should be sought.

Assume we have derived the following facts for one eight-hour shift which finishes 35,000 units. First, costs must be allocated between fixed and variable categories as described previously.

FINISHING FIRM A
COST ANALYSIS
At a Volume of 35,000 Units

Cost Factor	Variable	Mixed	Fixed
Materials	$ 886		
Chemicals	940		
Labor (wages)	7,692		
Polish	419		
Water	101		
Freight	117		

FINISHING FIRM A
COST ANALYSIS *(Continued)*
At a Volume of 35,000 Units

Cost Factor	Variable	Mixed	Fixed
Sales Tax	8		
Heat		$ 653	
Officers' salaries			$1,560
Social security taxes		542	
State unemployment taxes		353	
Federal unemployment taxes		37	
Fringe benefits		712	
Maintenance and repair		1,450	
Electricity		660	
Personnel insurance			55
Business insurance			426
Real estate taxes			293
Telephone			169
Dues and subscriptions			63
Business taxes			17
Depreciation			728
Vehicle and sales expenses		143	
Office supplies		29	
Professional fees			740
Totals	$10,163	$4,579	$4,051
*Distribution of mixed costs	+ 3,053	−4,579	+1,526
	$13,216	-0-	$5,577

* The mixed costs must be allocated between fixed and variable as previously described. For example, social security taxes and state and federal unemployment taxes will be fixed to the extent they are applicable to officers' salaries, and variable to the extent they are applicable to labor wages, unless some laborers earn more than the maximum amount on which social security taxes are based. To that extent, they would be fixed. Other mixed costs can be allocated using scatter diagrams, mathematical computations, or based on the experience of the manager, if he feels his observations have been sufficient to make reasonable allocations.

This analysis provides the following information:

Variable costs = $.3776 per unit ($13,216/35,000)

Fixed costs = $5,577

Further assume the selling price per unit is $.55. Then, the contribution margin per unit is $.17, calculated as follows:

Selling price per unit	=	$.5500
Variable costs per unit	=	.3776
Contribution margin per unit		$.1724

The contribution margin percentage per unit is 31.35 percent ($.1724/$.55). Consequently, the break-even point in sales is $17,790 ($5,577/31.35%) or 32,346 units ($17,790/$.55).

Proof:

Sales – 32,346 × $.55	=	$17,790
Variable costs – 32,346 × $.3776	=	12,213
Contribution margin		$ 5,577
Fixed costs		5,577
Profit before taxes		-0-

Considering the first eight-hour shift, we see the following results:

First 8-hour shift	
Sales 35,000 @ $.55	$19,250
Variable costs 35,000 @ $.3776	13,216
Contribution margin	$ 6,034
Fixed costs	5,577
Profit before taxes	$ 457

Now assume a second shift is added, with identical items being finished at the same selling price per unit.

Both Shifts	
Sales 70,000 @ $.55	$38,500
Variable costs 70,000 @ $.3776	26,432
Contribution margin	$12,068
Fixed costs	5,577
Profit before taxes	$ 6,491

This example makes it clear that profit before taxes increased a little over 14 times with a doubling of units produced and sold (adding the second shift). Since we know the way variable and fixed costs behave as volume changes, the conclusion can be reached very easily without the second example above. We were barely beyond the break-even point in the first shift. Therefore, any decision which increases production and sales will increase profit before taxes equal to the contribution margin (6,034 in this case). Therefore, adding a second shift will increase profits before taxes from \$457 to \$6,491. Another way to arrive at the same conclusion is that since 35,000 additional units will be finished, and since the contribution margin per unit is \$.1724, then profits before taxes will increase \$6,034 (35,000 × \$.1724) by adding the second shift.

Another way of coming to the same conclusion is to recognize that fixed costs are being spread over a larger number of units. For one shift, fixed costs per unit equal \$.159343 (\$5,577/35,000). For two shifts, fixed costs per unit dropped to \$.079671 (\$5,577/70,000). Variable costs, on the other hand, remain constant at \$.3776 per unit. Therefore, holding other things constant, the larger the number of units finished, the higher the *average* profit per unit will be, because fixed costs per unit will decrease while contribution margin per unit remains constant. Profit before taxes per unit will increase from \$.013057 (\$.172400 minus \$.159343) on 35,000 units (which equals \$457), to \$.092729 (\$.172400 minus \$.07967) on 70,000 units (which equals \$6,491).

Assume the same facts, except overtime premiums, additional social security taxes, retirement benefits, etc., for the second shift increase variable costs to \$.52 per unit. Since these costs change between alternatives, they are incremental and must be considered. The first 8-hour shift data above would remain the same as shown. However, both shifts combined would now appear as follows:

Both Shifts		
Sales 70,000 @ \$.55		\$38,500
Variable costs: 35,000 @ \$.3776	\$13,216	
35,000 @ \$.52	18,200	31,416
Contribution margin		\$ 7,084
Fixed costs		5,577
Profit before taxes		\$ 1,507

Since profit before taxes still increases from \$457 to \$1,507, a total of \$1,050, the second shift would still be added. Since the firm had passed break-even, there is a contribution margin of \$.03 per unit (\$.55 selling price less \$.52 variable costs) on the additional 35,000 units produced and sold, or a total increase in profits of \$1,015 (35,000 × \$.03).

Assume all the same facts in the preceding example, except fixed costs also increase by \$1,200. Since fixed costs change between alternatives (one shift vs. two shifts) they also are incremental in this example and must be considered as follows:

Sales 70,000 @ \$.55		\$38,500
Variable costs: 35,000 @ \$.3776	\$13,216	
35,000 @ \$.52	18,200	31,416
Contribution margin		\$ 7,084

Fixed costs: One shift	$ 5,577	
Increase in adding 2nd shift	1,200	6,777
Profit before taxes		$ 307

Since profits before taxes drop from $457 to $307, the second shift should not be added. This example also makes clear that incremental costs are not always only variable costs. If a cost changes between alternative choices, it is incremental and must be considered if a correct choice is to result.

One reason more mechanized finishing companies in this survey did not operate on a multi-shift basis is that the added productive capacity requires still more sales effort. Many companies obtain enough business for single shift operation with little outside customer contact. Doubling or tripling their sales levels for a two- or three-shift operation would certainly require adding one or perhaps two field salesmen.

Still, as noted in Chapter II, many successful survey companies point out that their around-the-clock operation actually attracted sales they otherwise might not get. Some highly motivated finishers picked up customers' priority work after their first shift, finished it during the night, and delivered it back to the customers before the first shift on the following day. This quick reaction capability enticed many customers to send *all* their finishing to a company having this capability — knowing that in an emergency, they can depend on them to do overnight work.

Besides placing increased demands upon the sales function, around-the-clock operation demands that the second- and third-shift foremen be highly qualified. Commonly, they alone are responsible for the entire plant after the owner-manager leaves for the day. Thus, production quality and decisions of the second- and third-shift foremen over the entire range of operations are an important factor in determining whether second and third shifts are profitable.

These are relatively simple examples, but they do indicate the types of knowledgeable decisions which can be made by proper use of cost data broken down into fixed and variable components. Such data can be utilized to aid in determining whether to add or drop a customer, to add or drop a particular type of finishing operation, to expand one type of finishing and reduce another, to accept one-shot orders at less than normal selling price where excess capacity exists, to convert to automated equipment, and to aid in pricing decisions, as discussed later.

A variable/fixed cost breakdown also allows a different presentation of income statement data from the conventional form presented in Chapter III. An example follows. Of course, the fixed and variable costs would have to be divided into manufacturing, selling, and administrative and general before the following statement could be prepared.

This should cause little difficulty, since most costs probably are initially classified into those categories for regular income statement purposes. Here we simply used the initial figures given in the preceding example and assumed the following classification into manufacturing, selling, and administrative and general costs.

Example I

METAL FINISHING FIRM

INCOME STATEMENT

For Month (year, quarter) Ending ______

		Dollars	Per cent of sales
Sales		$19,250	100.0%
Variable costs:			
Manufacturing	$8,156		
Selling	1,800		
Administrative and general	3,260	13,216	68.7%
Contribution Margin		$ 6,034	31.3%
Fixed costs:			
Manufacturing	$3,135		
Selling	300		
Administrative and general	2,142	5,577	28.9
Profit before taxes		$ 457	2.4%
Taxes		121	.6
Net Income		$ 336	1.8%

The statement above, referred to as a direct, variable, or marginal costing statement should be prepared in addition to the conventional income statement. The conventional income statement can be used as indicated in Chapter III. The statement here presents data in a different format, useful in different ways.

If more than one finishing line exists, the statement should be further refined as follows:

Example II

METAL FINISHING FIRM
INCOME STATEMENT
For Month (year, quarter) Ending ______

	Finishing Line X	Finishing Line Y	Finishing Line Z	Total
Sales	XX	XX	XX	XX
Variable Costs	XX	XX	XX	XX
Contribution Margin	XX	XX	XX	XX
Fixed Costs				XX
Profit before taxes				XX
Taxes				XX
Net Income				XX

The statement can reflect as much detail as management finds useful. Sales, variable costs, and contribution margins should also be reflected in dollar amounts per unit finished. A number of advantages accrue from such a statement:

1. By type of finishing operation, management can quickly see the impact of changes in sales on profits before taxes. Also, it is easier to observe what is happening to the profit picture because selling prices are changing and/or variable costs are changing, and to take action to correct problems which arise.
2. Management can see the contribution margin by finishing process, in total and per unit. Decisions can be more easily made to substitute one customer for another, accept one job as opposed to another, or in other ways change the finishing mix of work done, since whichever decision maximizes total contribution margin will maximize profit before taxes.
3. It stresses the lump sum amount of fixed costs to be recouped before income emerges. This helps to attract management's attention to fixed cost behavior and control when both short-run and long-run plans are being made.
4. It aids in the pricing decisions to be made by making it easier to observe which operations are carrying the major weight of profit generation, and which are loss operations where selling prices must either be increased or the operation dropped.

Similar data can be accumulated by major customers and/or major industries served, and they should be collected if either represents a material amount of the total sales activity of a finishing firm. The manager will then have information with the same advantages as those stated, only segmented by customers and/or industries served instead of by finishing lines (or in addition to segmentation by finishing line).

The ratios of fixed to total expenses and variable to total expenses are helpful in understanding cost-volume-profit relationships. In our example, this relationship is $5,577/$18,793 = 27.7 per cent for a sales level of $19,250. Should sales increase to $25,000, for example, this percentage should drop to 24.5 percent. [At $25,000 sales, variable expenses should be $17,175 ($25,000 times 68.7 percent). $17,175 plus $5,577 = total expenses of $22,752. Then, $5,577/$22,752 = 24.5 percent.] Thus, assuming some unused capacity, if sales increase, variable costs as a percentage of sales should stay constant at 68.7 percent, but should increase as a percent of total expenses. Fixed costs should decrease, both as a percent of sales and as a percent of total expenses.

If such patterns do not follow changes in sales volume, management should look behind the scenes to ascertain the reasons. Taxes, insurance, and other costs fixed per period may be increasing significantly between periods. Variable costs may be increasing materially. Both must be recovered in increased sales prices if profit margins are to be maintained. Volume of activity might be falling off, indicating a need to increase or redirect sales activity. The firm may be utilizing cost factors (materials, labor, overhead) less efficiently, requiring greater emphasis on cost control procedures and techniques.

Conclusions

Fixed and variable cost classifications and related analyses can be very important to the assurance of profitable operations in a finishing company. The industry, with its large proportion of relatively small operations, is presently not utilizing these analyses nearly to the extent that might be done to improve its efficiency and effectiveness. It is encouraging to note though that there has been a marked increase in the use of such data since 1967, as reflected by responses to our questionnaire. We applaud the progress, but a great deal more still needs to be done.

Viewing potential profitability in a setting which recognizes the nature of cost behavior is essential to even the smallest operator and is well within his reach. The approach is not complicated or difficult to understand. It improves the manager's ability to think of the total effects of his many important business decisions.

There are still too many accounting systems and reporting practices in use in the finishing industry which do not involve either separation of costs into fixed and variable groups or any recognition of cost behavior patterns. Accounting practices which may be developed to facilitate reports of a conventional nature and the preparation of tax returns should not be allowed to limit the effectiveness of the accounting or information function in a finishing operation, regardless of size.

The use of fixed and variable cost classification also is most important in answering the make-or-buy decisions of those who have established captive shops, as we discuss in Chapter VIII.

V

FINANCIAL STATEMENT ANALYSIS AND INTERPRETATION

The financial statements presented in Chapter III (balance sheet, income statement, funds statement) contain much useful information. For example, when one looks at the income statement, a net income figure tells a different story about the firm than a net loss; on the balance sheet, an overdraft at the bank gives the reader a different feeling as to the financial condition of the firm than a sizable amount of cash in the bank. Footnotes to the statements give a considerable amount of additional information useful in understanding the firm. Countless other examples could be given to show that financial statements, taken by themselves in the form normally presented, do convey useful information. To understand the meaning conveyed by the statements, however, — to understand fully the stories they tell about the firm — much more is needed. This "much more" is financial statement analysis and interpretation.

Financial statement analysis doesn't *add* data to those encompassed in the basic statements, but it does bring data to the surface in a more understandable form. Once the data are clearly disclosed by analysis, interpretation is concerned with accurately assessing the meaning of the data so disclosed.

Analysis can take on many forms. Perhaps the most useful, and the one we emphasize here, is ratio analysis. Although sixty-seven percent of questionnaire respondents indicated they are using ratio analysis in order to understand their business operations better, our field interviews indicated that the meanings conveyed by ratios frequently are not fully understood nor utilized in the most effective way. In other words, even though analysis is common, interpretation and use of the analysis is weak.

A ratio is simply a mathematical association of one thing with something else, which provides significant information for making decisions about the firm. From a financial viewpoint, it can involve the relationship of two items in the balance sheet, two items in the income statement, two items in the funds statement, one or more items from one of these statements with one or more items from one of the other statements. It is a meaningful relationship of certain items designed to bring to the fore information about the firm not apparent on the face of the statements, taken alone. Ratios indicate certain characteristics about the firm's financial structure that are not revealed by dollar comparisons — characteristics necessary to determine the financial position and operating results of a firm, including meaningful trends over the years — as well as to compare different firms in a meaningful way. For example, assume the following data for firms A and B.

	A	B
Sales	\$1,000,000	\$100,000
Net Income	\$100,000	\$10,000

It is obvious that A's sales and net income are ten times larger than those of B. Taken alone these relationships might be significant. For example, if the two firms operate in the same market, quite obviously A has a much larger share of the market than B. This is useful information. In terms of relative efficiency in the production of income, however, the firms are identical, since the percentage of net income to sales (net income divided by sales) is the same for each firm

$$\frac{\$100,000}{\$1,000,000} = 10 \text{ percent for A;}$$

$$\frac{\$10,000}{\$100,000} = 10 \text{ percent for B.}$$

Thus, this ratio reveals meaningful data not apparent from comparison of dollar figures alone.

Classification and Uses of Financial Ratios

There are four broad classifications of financial ratios: liquidity, leverage, activity, and profitability ratios. Specific ratios under each of these classifications will be presented and discussed fully later in this chapter. First we feel it worthwhile to indicate briefly the nature of the broad classifications of ratios, who uses them, and generally how they ought and ought not be used.

Liquidity ratios measure the ability of the firm to meet its maturing obligations and, hence, are important in evaluating the financial position of a firm. The largest cause of business failure is still the inability to meet obligations as they mature. Understanding such ratios, therefore, is critical for the metal finishing manager if he is to operate successfully. Such ratios are also considered carefully by bankers and other credit lenders before making funds available to a firm. The three main ratios for determining liquidity are the current ratio, the quick or acid-test ratio, and the defensive interval ratio.

Leverage ratios measure the investment of the owners (equity investors) vis-a-vis that of the firm's creditors, as well as the ability of the firm to cover fixed charges (such as interest). Such ratios indicate financial leverage. As we see later, financing the firm's operations to a reasonable degree with debt is good. Too much debt, on the other hand, might indicate future difficulties for the firm in terms of paying off the debt as it matures. An optimum relationship is needed, and the specific ratios which indicate this are debt to total assets, times interest earned, and cash flow per share.

Operating leverage, concerned with break-even analysis and the percentage of fixed to total expenses, is also an important consideration. These relationships were considered in Chapter IV.

Activity ratios indicate the efficiency and effectiveness in the use of specific assets, asset categories, and total assets in carrying out the objectives of the firm. As previously discussed, the more rapidly assets are turned over in the business, individually and in total, the better (holding other things constant). Assets which turn too slowly might indicate too large an investment in those assets, considering the sales and profits they generate. Or it might indicate additional sales effort is needed. Be assured, in any case, that firm managers must be alert to ways of utilizing assets more productively if the firm is to reach the maximum level of profitability. Creditors also keep a watchful eye on asset turnover ratios. Major examples of ratios in this category are receivable turnover, inventory turnover, and total asset turnover.

Profitability ratios measure the overall efficiency and effectiveness of the firm in attaining its primary objective. They help in answering such questions as: Was the profit what it should have been? What rate of return does it represent? Major examples of ratios falling in this category are profit margin on sales, rate of return on assets employed, rate of return on stockholders' equity, and earnings per share.

Who uses such ratios? The answer, of course, is anyone who needs to understand the firm well enough to make informed decisions. More specifically, the main users fall into two broad categories: outside creditors and investors comprise one of these groups, while internal management makes up the second. Although our main concern in this monograph is with internal management, finishing managers do need to understand which ratios others use and how they make use of them. For example, in obtaining a bank loan, it would be helpful for managers to know how bankers use ratios, what they look for, and how they evaluate what they see.

Ratios should not be taken at face value, and more is needed for proper interpretation than ratios from only one set of statements. Trends in ratios are important. Comparisons with "norms" for the industry are important. And looking behind the scenes to make sure the ratio really indicates what it purports to indicate is important. For example, observing that the income for the firm for a year is, say, $10,000, is important. The quantity, or dollar amount, is significant. And, when considering income the larger the dollar amount, the better. More important than just the dollar amount, however, are: the trend in income for the firm, the quality or source of the income, and the comparison of the income of this firm with others in the industry and with norms for the industry. For example, what if the $10,000 net income was the result of an extraordinary, unusual gain of $50,000 (such as might result from fire destruction of an insured building), offsetting a $40,000 operating loss. Such a net income figure takes on a decidedly different character than if had it all resulted from the normal, recurring operations of the firm. Or, let's assume the $10,000 is 2 percent of net sales. What if other firms in the industry are presently earning at a rate of 10 percent of net sales? Or, what if the "normal" earning rate for the industry is 12 percent? Comparing the firm's 2 percent with these figures gives additional information to judge performance of the firm. Holding other things equal, our example firm is not performing very well.

One last brief example to drive home the general points being made here. What if we determine the percentage of net income to sales ratio for each of the past five years, and find these have increased from a 15 percent loss on sales to the 2 percent given above. Assume further that present management took over the firm five years ago. Now a different story is told. Even though present performance is not what it ought to be, the trend certainly is in the right direction. It appears (considering only this one ratio alone) that management has been doing an effective job in bringing the firm back. If the trend continues we can expect the firm to reach the level of other firms in the industry and the "norm" for the industry, perhaps even to surpass both. In sum, trends in ratios, comparison of the firm's ratios with other firms in the industry and "norms" for the industry, looking behind the scenes at the specific items making up the categories used for calculating the ratio, all are important if correct interpretations are to be made. And unless correct interpretations are made, decisions based on those interpretations are not likely to be right ones.

But what is meant by "normal" ratios for the industry? A normal ratio for an industry is a relationship that should exist, assuming normal economic and financial conditions in the economy in general and in the particular industry specifically. A given ratio size is sometimes "normal" for most, if not all, industries. For example, a current ratio of 2 to 1 is usually considered normal for most industries. Other ratios, such as debt to equity, are considered normal at one level for a given industry, and normal at quite a different level for another industry. Does this mean that things are necessarily bad if a firm has some ratios out of line with norms for that industry? It might, but not necessarily so. Remember, normal ratios have evolved over the years based on assumed normal economic and financial conditions in the economy in general, and in the industry specifically. If conditions are not normal (high inflation rate, tight money, etc.) a firm's ratios may be out of line with norms and still be perfectly acceptable. In fact, such conditions might cause an entire industry to be out of line with the norms. Or management may make purposeful, knowledgeable decisions which might cause some of its ratios to be temporarily out of line with the norms for good and sufficient reasons.

Normal ratios, however, do act as standards for comparison. When a firm's ratios deviate from these norms, then management should find out why. A good, acceptable explanation (reason) may result, or the difference might indicate a problem needing correction. We will emphasize such analyses as we consider each ratio specifically. Although normal ratios are published for many industries, we know of no publication listing normal ratios specifically applicable to metal finishing. In Chapter VIII we recommend that such "norms" be established by the National Association of Metal Finishers, and we indicate how to go about accomplishing this goal. More industrial data are needed for metal finishing firms.

Throughout the following specific examples, we use the data from the example financial statements presented in Chapter III.

Liquidity Ratios

Liquidity ratios are intended to reflect the debt-paying ability of a firm; i.e., whether or not it can pay its current obligations as they come due.

The *current ratio* is determined by dividing current assets by current liabilities. The current ratio would be

$$\frac{\$58{,}000}{\$39{,}800} = 1.46 \text{ to } 1,$$

meaning simply that the firm in question has \$1.46 of current assets on hand for every \$1 of current liabilities owed. Is this adequate? Does it indicate the firm will be able to pay its liabilities as they come due? A number of other considerations are necessary in order to answer these questions? The "normal" current ratio is deemed to be 2 to 1, so the firm in question certainly is below that figure. But that norm simply indicates that under normal economic and financial conditions, a 2 to 1 margin indicates a coverage adequate to avoid short-term difficulties. Anything less than 2 to 1 means the firm is accepting more financial risk than normal. But in inflationary periods management may decide deliberately to keep the current ratio below 2 to 1 in order to avoid some of the loss in purchasing power of dollars held in a checking account or invested in idle inventory. In addition, there is an opportunity cost associated with any asset which is idle, such as cash or inventory; that is what the dollars tied up in such assets could be earning if placed in alternative investment opportunities, such as a savings account or marketable securities.

Reducing such opportunity costs and losses in purchasing power by keeping assets turning rapidly is a good management decision, *provided* funds are available to pay debts when they come due. The lower the current ratio, the less likely such funds will be available, and the greater the financial risk assumed by management. Obviously, the ultimate decision rests on a balancing of such factors.

If we assume that inventories will be converted into finished products or services, that these will be transferred to the customer and the accounts receivable so generated (plus those presently in existence) will be collected in the normal course of business, then the 1.46 to 1 ratio may be adequate. But what if the only significant item in current assets was inventory (not the case in our example, of course), and what if inventory turned slowly? Then a 1.46 to 1 current ratio would be inadequate and would portend serious financial difficulties for the firm.

This implies a number of different things. First a normal ratio (in this case 2 to 1 for the current ratio) assumes a normal relationship among specific items in the categories used to calculate the ratio. For the current ratio, it implies a normal relationship between cash, accounts receivable, inventories, and prepayments in the current assets (numerator), and a normal relationship between the specific items in current liabilities (denominator). What these normal relationships should be depends upon the activities and objectives of the particular firm, but, in general, cash must be adequate to take care of the day-to-day operating needs of the firm. Accounts receivable must be maintained at an optimum level. For example, having zero accounts receivable generally would not be good since such a condition probably would only exist if the firm sold only for cash. If it did, total sales would be lower than if it sold on account. On the other hand, too large a balance in accounts receivable might indicate a slow turnover; i.e., an inability or difficulty in collecting accounts in accordance with credit terms. Essentially the same points could be made regarding inventories. The essential point is that management must maintain the optimum amounts in specific asset and liability categories. This is difficult, but necessary if long-range profits are to be maximized.

If 2 to 1 is good, does this mean 4 to 1 is better, 10 to 1 is better still, etc.? In some cases, the larger the ratio the better. This is not true for the current ratio, however. If 2 to 1 is normal (and remember what the implications of "normal" are), the larger the ratio becomes, the greater the amount of assets which are idle and the greater the opportunity costs and losses in purchasing power in inflationary times. However, management might decide to increase inventories substantially above normal levels (and, depending on the circumstances, the current ratio) because of an anticipated severe price rise for materials. If the price rise does come about, and assuming the savings resulting from stocking up exceed the opportunity costs associated with the capital tied up in inventories, then this would be a good decision by management and would explain why the current ratio exceeds 2 to 1. If the price rise does not materialize, then the decision was a bad one.

Finally, trend lines are important. What if the current ratios for the past five years are considered, and we find a steady decrease from 2.3 to 1 to 1.46 to 1? Looking behind the scenes we might find the decrease resulted from a deliberate and sound management decision, such as those indicated above, and nothing alarming is indicated. On the other hand, we might find, for example, that funds generated by operations have been decreasing substantially because increases in selling prices have not kept pace with increases in cost factors. If this is true, the situation will have to be reversed or the firm might face substantial difficulties in paying its obligations as they come due. Selling prices will have to rise and/or the prices of the cost factors will have to be controlled better.

In summary, the current ratio should not be taken at face value, as it so often is. A look behind the scenes is necessary. Various comparisons must be made and trends must be considered if the meaning of the ratio is to be interpreted properly in the circumstances and, as a consequence, if correct decisions are to be made based on it. The same observations are true for all other specific ratios we will cover, but, in the interest of space, we will not repeat them in all cases.

Another important liquidity ratio is the *quick or acid-test ratio.* This is calculated by dividing cash plus net accounts receivable (and marketable securities, if any) by current liabilities. Using our example data,

$$\frac{\$42{,}000}{\$39{,}800} = 1.06 \text{ to } 1.$$

The normal for this ratio is 1 to 1. It is a more severe test of debt paying ability since inventories and prepaid expenses are excluded from the numerator. Since inventories and prepayments are further removed from cash than accounts receivable, many feel this is a better test of debt paying ability than the current ratio. There is some question as to how long it will take to convert inventories into finished goods or services and how much might be realized on their sale; so excluding them is warranted.

The 1.06 to 1 in our example, which is somewhat larger than normal, might indicate that accounts receivable are turning too slowly in terms of the credit terms allowed. Perhaps the estimate for bad debts is not high enough, leaving net receivables stated too high. In other words, when this ratio is too large, it might indicate some problems with receivables or signify that current liabilities are very low. One might think a low total for current liabilities is good, but this is not necessarily so. It might indicate an unfavorable, or at least less than an optimal, relationship between current liabilities, long-term liabilities, and stockholder's equity. Although we discuss such relationships more fully later, remember that current liabilities generally are a cheap source of funds. Interest rates on short-term obligations usually are lower than on long-term debt, and credit supplied by trade creditors is often easily obtainable and relatively inexpensive.

Obviously, it does not follow from these assertions that we should have all short-term debt and zero long-term debt. It also does not follow that we should have zero cash and receivables and large amounts of current debt. As the quick ratio drops below 1 to 1, more and more financial risk is being assumed by the firm. If a point is reached where debts cannot be paid as due, the firm may be driven into insolvency and ultimately into bankruptcy. The discussion does indicate, however, that management must be concerned with the optimum relationships between specific items in the categories used in calculating the ratio, as well as the optimum relationship between the categories themselves.

The *defensive interval ratio* is relatively new and is not often used in the finishing industry. Interpreted properly, however, it also can provide useful information about the liquidity of the firm. The defensive interval ratio is determined by dividing cash and net accounts receivable by projected daily operational expenditures. Using the example data, the numerator would be $42,000. The denominator would be production expenses (cost of goods sold), plus selling expenses, general and administrative expenses, and taxes, minus non-cash expenses, such as depreciation, divided by 365 days. Assuming total depreciation expense is $17,000 in our example, the denominator would be $633,607, calculated as follows:

Production expenses	$449,181
Selling expenses	24,296
General and Administrative expenses	156,055
Taxes	21,075
	$650,607
Less: Depreciation	17,000
	$633,607

The defensive interval ratio therefore is

$$\$42{,}000 \div \frac{\$633{,}607}{365} = 24.19 \text{ days}$$

The ratio indicates the firm's capability to meet its basic operational costs. In this case, the "defensive" assets will cover about 24.19 days of operational needs. Other factors such as credit terms on sales and purchases, as well as timing of cash inflows and outflows, would need to be considered before a knowledgeable decision as to the adequacy of 24.19 days could be judged. As a general rule, however, such a low coverage would be cutting it pretty fine. Usually 30 days coverage of operational needs is the minimum needed for adequate protection for the firm.

Activity Ratios

Activity ratios indicate how efficiently and effectively the firm uses its assets. Holding other things constant, the more rapidly assets are turned over, the greater the profit generated. This is true for specific assets such as accounts receivable and inventories; it is true for categories of assets such as current assets; and it is true for assets in total.

Accounts receivable turnover is calculated by dividing net sales on account by average net receivables. Two things must be emphasized. The numerator should not include cash sales (if cash sales remain relatively constant, or are small in amount, little harm is done by including them). The denominator should be determined by adding beginning of the period net accounts receivable to the end of the period net accounts receivable and dividing by two. Using the data in our financial statement examples, assuming all the sales were on account, and that beginning of the year accounts receivable (net) amounted to \$33,000, receivable turnover would be

$$\frac{\$709{,}607}{\$\ 34{,}000} = 20.87$$

times per year. Dividing 365 days by 20.87 equals 17.49 days; i.e., receivables are being collected every 17.49 days on the average.

To determine whether this is good, bad, or indifferent, other things must be considered. What if the firm's terms of sale require payment in five days? This would indicate that a large percentage of the receivables are past due. This might further indicate that credit was extended too liberally — that the firm accepted a number of bad customers because their credit standing was not checked closely enough. It might indicate that the firm's collection procedures are too lax. Either situation is undesirable. Slow collections can create cash shortages, making it difficult to pay liabilities as they come due. Large bad debt losses may occur, reducing the income of the firm.

Assume, instead, that the firm's credit terms were for thirty days. Would this mean that the average collection period of 17.49 days is good? It might, assuming all other conditions are normal. But it might indicate an excessively restrictive credit policy which limits sales only to customers of the highest credit rating. If the income that is lost because sales are not made to customers of lower quality exceeds the bad debt losses that would result by expanding sales to that group, then the credit policy is too restrictive and profits are being lost to the firm. Comparisons of a firm's average collection period with the industry average and norms for the industry can be used to refute or support such conclusions. Trend lines are also important to judge the direction of movement of receivable turnover. Material changes need to be investigated.

Inventory turnover is determined by dividing cost of goods sold by average inventory (beginning of the period inventory plus end of the period inventory divided by two). Assuming beginning of the year inventory was $13,000, then inventory turnover is

$$\frac{\$449,181}{\$\ 12,000} = 37.43$$

times a year. This equals a turnover of once every 9.75 days, simply calculated by dividing days of the year (365) by 37.43.

Generally speaking, the more rapidly inventory turns the better, holding other things constant. The larger the number of units a firm is able to sell, assuming the gross profit margin per unit and the average inventory level remain constant, the higher the firm's profit. Whether the trend is toward a more rapid, or a less rapid, turnover, however, additional investigation is needed to determine whether such a trend is good or bad. Sales price reductions might speed up the turnover while reducing profits. An active marketing program might be undertaken at considerable expense in order to increase sales, and this might increase turnover. However, income would decrease if the expenditures for the program exceeded the additional profit generated by it.

A low ratio, on the other hand, might indicate obsolete inventory or stockpiling too much inventory. As we have stated, inventories must be maintained at an optimum level, sufficient to meet production requirements as needed; but no more in excess of that amount should be stocked than is needed as a safety factor. Investment in inventory above such amounts represents idle capital, with the attendant opportunity costs involved. A low turnover might also indicate repricing problems that need attention. In any case, as is true for other ratios, the size of the ratio taken by itself means but little. Factors backing the ratio must be considered, trend lines determined and analyzed, and comparisons made with industry averages and norms, if the ratio's full significance is to be understood.

We can now start to see some of the interrelationships and interdependencies between the ratios; i.e., in many cases, the complete meaning of one ratio cannot be understood and properly interpreted without considering one or more other ratios. For example a weak current ratio can be offset by high turnover of accounts receivable and of inventory. Even a low acid-test ratio can be strengthened by rapid turnover of accounts receivable and of inventory.

Asset turnover is computed by dividing net sales by average total assets (assets at the beginning of the period plus assets at the end of the period divided by two). For our representative company, the asset turnover (assuming total assets at the beginning of the year were $360,000) would be

$$\frac{\$709{,}607}{\$366{,}500} = 1.94 \text{ times per year.}$$

This ratio, particularly when compared with industry averages and norms, gives an indication of how efficiently the management of a particular firm utilizes its assets compared with other managers in the industry. Ideally, in computing this ratio, current replacement costs for assets should be utilized rather than book values, although this is rarely done. Comparisons with other companies using current replacement costs would be more meaningful. Using book values (the usual case), the impression is given that firms using old assets (low book values) are more efficient than firms using new assets, which may not be the case at all. Consequently, the ratio must be used carefully and in conjunction with other ratios which indicate or measure efficiency if it is to be properly interpreted. Nevertheless, properly considered, it can be useful.

Leverage Ratios

Leverage ratios aid in predicting the long-run solvency of the firm. They are helpful in determining the optimal capitalization structure of the firm; i.e., deciding how much debt should be utilized relative to stockholder's equity, and what is the optimal relationship between current liabilities and long-term debt.

One might be tempted to conclude that the best capital structure in all circumstances would be 100 percent stockholders' equity — a situation where all of the assets of the firm are provided by the owners. This would only be true if a serious depression were on the horizon, or if the firm could earn less by using the assets in its operations than the amount that it would have to pay the creditors for use of their funds (net of taxes). Since common stock does not have a fixed cost or a fixed maturity date, as long-term debt and short-term debt do, it would be best for the firm to have zero liabilities when a serious depression started. No one could demand payment of interest or principal and drive the firm to bankruptcy in the event of default. The firm could simply perform minimal maintenance work and await renewed demand for its product or service to start operating again. Or, if a firm is able to earn less on borrowed funds than the interest payments (net of taxes) for those funds, then profits would be lower or losses larger when using borrowed funds as opposed to obtaining those funds from the owners. But these are somewhat unusual circumstances.

If economic and financial circumstances are relatively normal, many firms can earn a rate of return on assets employed which exceeds the cost of borrowing (net of taxes). It is to the firm's advantage to borrow in such circumstances, since the amount earned in excess of the cost will increase the profit to the owners without their risking any additional funds in the firm. This is referred to as "trading on the equity." For example, if the current borrowing rate is 8 percent, the effective tax rate is 40 percent and the firm can earn at a 7 percent rate of return on total assets, a profit advantage would accrue to the owners if the funds were borrowed rather than supplied by the owners. The net cost of borrowing would be 4.8 percent (8 percent less 40 percent × 8 percent). This is 2.2 percent less than the 7 percent earning rate. This difference leaves an additional margin for the owners without their risking even $1 of additional investment in the firm. If the rate of return on stockholders' (owners') equity is higher than the rate of return on total assets employed, this indicates the firm is trading on the equity successfully.

Does it follow from this, on the other hand, that virtually all funds needed by the firm should be supplied by creditors? Absolutely not. The larger the debt to equity ratio, the greater the financial risk assumed by the firm. A point is reached where the additional risk of issuing debt securities exceeds the additional benefits of larger profits for the owners. What that cost/benefit relationship is cannot be determined with exactitude. It depends partially, for example, on management's willingness to accept risk. Some managers are more conservative than others. It also depends on the nature of the demand for the product or service produced. It the demand for the product or service is relatively inelastic, the consumer will continue to buy the same quantity even in the face of relatively large price changes. Under such conditions, the revenue of the firm will stay relatively constant, and, assuming out-of-pocket expenses are kept under control, funds generated by operations will stay fairly constant. Since these funds can be utilized to make interest and principal payments, larger amounts of debt can be carried because of the dependability of the funds generated.

The opposite conditions prevail when the demand for the product or service is relatively elastic. Small changes in price will generate relatively large changes in the product or service demand, causing funds generated by operations to fluctuate widely. Therefore, the risk is greater that sufficient funds will not be available to service the debt, so smaller amounts of debt can be safely carried.

Consequently, a firm such as a public utility, where demand is relatively inelastic, usually can carry a larger percentage of debt to equity and assume no more risk than, for example, an automotive producer, where the demand is more elastic for the product produced and sold. Thus, when interpreting the following ratios, in addition to the other comparisons we have discussed for other ratios, the manager must keep in mind the nature of the demand for the product or service his firm provides. Metal finishers, tied as they are to the business cycles of their customers, must consider the demand characteristics of the products provided by their customers when deciding upon the optimal debt to equity ratio for their firms.

The *debt to total asset ratio* is calculated as follows:

$$\frac{\$159{,}800}{\$373{,}000} = 42.8 \text{ percent.}$$

This indicates that 42.8 percent of the assets needed by the firm are provided by the creditors. The other 57.2 percent are provided by the owners. If the industry average is somewhere near this figure, it probably indicates the firm is making effective use of debt, and that the financial risk assumed is within the bounds of acceptability. As pointed out, however, it is even better to compare this ratio with only those firms in the finishing industry performing the same types of services for the same industry groups, since the demand characteristics of products provided by the finishing firm's customers is an important consideration in deciding upon the "best" debt to equity ratio. For example, this ratio of 42.8 percent is close to those maintained by firms in the automotive industry.Thus, if this finishing firm is tied closely to automotive companies in the finishing it provides, and assuming the automotive companies are financed well, this is further evidence that the ratio is within acceptable bounds. If our later calculations show a return on stockholders' (owners') equity larger than the return on total assets employed, this will indicate successful trading on the equity by the firm.

The ratio is sometimes computed by dividing debt by owners' equity, as follows:

$$\frac{\$159{,}800}{\$213{,}200} = 74.9 \text{ percent.}$$

This ratio, properly interpreted, tells exactly the same story as the preceding calculation, but in a somewhat different way. This ratio says that for each $1 of assets provided by the owners, 74.9 cents of assets are provided by the creditors. The important thing is that the calculation, whichever one is used, be made the same each year for trend line purposes, and the same way as industry averages and norms are computed for comparison purposes. Otherwise, the trend lines and comparisons are worse than meaningless; they are misleading.

The *times interest earned ratio* is computed by dividing income before taxes and interest charges by interest charges. If we assume interest charges equal $9,000 in our income statement, and the effective tax rate for the firm is 25 percent, then the calculation would be

$$\frac{\$80{,}075 + (\$6{,}000 - 25\% \times \$6{,}000)}{\$9{,}000} = 9.40 \text{ times.}$$

The larger the times interest earned ratio, the more likely interest can be paid when due. Although it is based on an income figure, this calculation really is getting at the funds (generated by operations) available to cover interest charges when due. Consequently, a better calculation is to divide funds provided by operations before taxes and interest by interest charges, as follows:

$$\frac{[\$75{,}400 + \$9{,}000 + (\$21{,}075 - \$300)]}{\$9{,}000} = 11.69 \text{ times}$$

Again, the greater the elasticity of demand for the firm's product or service, the higher this ratio needs to be for safety, because of the large variations in funds provided by operations that might take place.

Cash flow per share is determined by dividing the cash provided by operations (from the funds statement) by the number of common shares outstanding. Assuming our firm has 16,000 shares of common stock outstanding, then the calculation would be:

$$\frac{\$75{,}400}{16{,}000} = \$4.71 \text{ per share.}$$

This ratio is not a substitute for any of the profitability ratios. It has nothing to do with determining how profitable or unprofitable the operations of the firm are. It simply indicates the cash per share generated by operations. This ratio, particularly the trend lines of cash flow per share which will indicate its degree of variability between years, is important in deciding how much debt can be carried safely. Payment of debt has a high order of priority when determining how the cash generated by operations will be utilized, certainly higher than the payment of dividends or using cash for expansion purposes. Consequently, it is a useful ratio in leverage or coverage ratio analyses.

Profitability Ratios

Profitability ratios give some indication of how successfully or unsuccessfully the firm has operated during the year. Profitability ratios are utilized as the ultimate test of managment effectiveness and efficiency. Usually, the higher these ratios, the better.

Profit margin on sales ratio is computed by dividing net income by net sales. In our example, this would be

$$\frac{\$\ 29{,}000}{\$709{,}607} = 4.1 \text{ percent.}$$

A better calculation is to divide income before extraordinary items by net sales, or

$$\frac{\$\ 59{,}000}{\$709{,}607} = 8.3 \text{ percent.}$$

This ratio is better, particularly when considering trend lines, since it reflects the rate of return on sales from the normal recurring operating activities of the firm. Consequently, it is more useful for predicting what is likely to happen in the future, and also for comparisons with industry averages, than one which includes the results of unusual nonrecurring extraneous events. If the industry average is 7 percent, for example, the 8.3 percent indicates that the firm is achieving an above average rate of return on each sales dollar. This ratio, taken by itself, does not indicate how profitable the firm was for a given period of time, in spite of the fact that most people think it does. Return on investment is the only "true" indication of profitability, and the profit margin on sales must be utilized along with the asset turnover ratio in order to indicate this. For example, many retail food chains have a net income-to-sales ratio of 1/2 of one percent. For a jewelry store it may be 7 percent, 14 times as high. Does this mean the jewelry store is 14 times as profitable? Not at all. Rate of return on assets is the critical indicator of profitability, and this must consider asset turnover.

Thus, the *rate of return on assets employed* equals the profit margin on sales times asset turnover: 8.3 percent × 1.94 = 16.10 percent. Another way of calculating the rate of return on assets employed is to divide income before extraordinary items by average total assets. Thus,

$$\frac{\$\ 59{,}000}{\$366{,}500} = 16.10 \text{ percent.}$$

Still another way, which shows the interrelationships the best, is

$$\frac{\text{income before extraordinary items}}{\text{net sales}} = \frac{\text{net sales}}{\text{total average assets}} =$$

$$\frac{\$\ 59{,}000}{\$709{,}607} \times \frac{\$709{,}607}{\$366{,}500} = 16.10 \text{ percent.}$$

The 16.10 percent would have to be compared with industry averages, target rates of return, and its trend line for a number of years in order to determine if it represents a rate of return at least equal to the risks assumed in the venture.

This ratio also brings to the fore again the importance of rapid asset turnover. Some firms, such as the large retail grocery stores referred to previously, frequently make only ½ to ¾ of one percent margin on sales, but their assets turn over so rapidly that this provides an attractive rate of return on assets employed. Other firms, such as jewelry stores selling high quality items, must have a 7 to 10 percent or higher margin on sales in order to stay in business, because items such as jewelry turn over so slowly.

Return on stockholders' (owners') equity is determined by dividing income before extraordinary items by average stockholders' equity (beginning of the period stockholders' equity plus end of the period stockholders' equity divided by two). Assuming beginning of the year stockholders' equity was $195,000, then

$$\frac{\$\ 59{,}000}{\$204{,}100} = 28.9 \text{ percent.}$$

Since the rate of return on total assets employed is lower than this figure, the firm is trading on the equity at a gain. In the calculation of this ratio, as well as all others making use of income statement data, a reasonable salary must be deducted for the owner-manager. One of the major purposes of profitability ratios is to indicate whether or not a sufficient rate of return on capital investment is being generated by the firm. Any return to labor (a reasonable wage) of the owner-manager must be deducted from "income" figures used, or the return will be overstated and wrong inferences derived.

Another profitability ratio frequently used is *earnings per share.*

$$\text{Earnings per share} = \frac{\text{income before extraordinary items}}{\text{shares outstanding}} =$$

$$\frac{\$\ 59{,}000}{16{,}000} = \$\ 3.69 \text{ per share.}$$

This figure, though useful, can be deceptive. For example, one would expect this figure to increase every year that the owner "plows back" (reinvests) his earnings into the business. The additional investment should generate larger profits and, if the number of shares outstanding remains constant, then earnings per share must increase. In no way does this indicate whether the return is sufficient under the circumstances, nor does such an increase reflect any increased efficiency of the firm in providing its product or service. In fact, looking behind the scenes might indicate the opposite. Nevertheless, when used properly, and when compared with its trend lines, other firms in the industry, and industry norms, this ratio does provide additional useful information.

The Funds Statement - Additional Considerations

In addition to those items already considered in this chapter and Chapter III, a few additional points need to be made regarding the use of the funds statement.

The funds statement provides data useful in obtaining loans, particularly those of an intermediate- and long-range nature. The following data in particular can be helpful to bankers and other credit suppliers of the firm, as well as to finishing firm managers, in analyzing and understanding the financial standing of the firm:

1. Cash flow from operations — interest and principal debt payments must be made primarily from funds generated by operations. Trend lines indicating stability in this figure, or better, increases over a period of time, can be helpful in obtaining loans.
2. Financing activities — indicates trends in obtaining outside financing; whether emphasis is on debt or equity financing; whether amounts borrowed are consistently exceeding repayments, i.e., whether debt is increasing and becoming unwieldly or is under control and being used wisely.
3. Funds utilized for expansion — gives creditors information on whether a firm is expanding and modernizing operations. If a firm is expanding and modernizing, and assuming rates of return have been consistently adequate, this indicates the firm is likely to continue successfully into the future long enough to repay interest and principal on the debt as it comes due, and is a positive factor when obtaining a loan.

Concluding Remarks

Properly utilized, ratios can aid management in operating the firm efficiently and effectively. But they should never be accepted at face value. Each one represents the compression of large amounts of financial data into one figure. Ratios cannot, therefore, be fully understood and interpreted unless there is reasonable understanding of the data on which they are based. For example, financial statements are not adjusted for price-level changes. Inflation or deflation can greatly affect financial data and the ratios based on them, depending upon such things as the type and age of the assets involved. These must be considered when interpreting the meaning of ratios.

Too much emphasis should not be placed on one or two ratios taken alone. A weakness in one can be counterbalanced by strengths in others.

Together, not alone, they tell the financial and operational story of the firm and should be considered, primarily, in this total setting. On the other hand, it would be possible to calculate scores and scores of ratios, i.e., one could relate almost any item in any statement to almost any other item in that statement or others. The question is whether such an approach would be helpful. The answer is "no". First of all, too many ratios would cause the user to get lost in a morass of detail. Second, and most important, a ratio should be the relationship of items having material significance to one another. Otherwise, the result is not meaningful and useful. It is not a question of whether the items can be casually related, but whether that relationship has material significance for analysis purposes. If not, the ratio should not be calculated and used.

Uncertainty exists throughout the accounting process — in the measurement of income, in the measurement of assets, and in a myriad of other ways. Such measurements are approximations at best — needed and useful — but approximations nonetheless. Consequently, ratios based on such data are approximations, and in no sense are they as accurate as the final calculations might imply. This does not mean they are not useful, only that they be considered with due care.

Neverthless, ratios are useful as one tool in analyzing a firm. Ratios other than those given might be calculated and used, and they often are. The ones discussed here are important ones, perhaps the most important, and discussing them can act as a springboard for calculating and understanding others, if they are considered needed and useful by firm managers.

VI

PRICING POLICY - ITS ESSENCE AND ECONOMIC UNDERPINNINGS

Pricing wisely is one of the most important functions for any successful firm, regardless of product produced or service provided. Certainly this is no less true for firms in the finishing industry. Each firm, and indeed entire industries, over the long run must make a rate of return on capital committed commensurate with the risks of the undertaking, or go out of business. To do this, many tasks must be performed well. One of these, certainly, is effective pricing of jobs and services. Weaknesses in pricing can cause losses and eventual business failure. Incorrect pricing policies of one or a small group of firms can have far-reaching, detrimental effects on large numbers of firms in a particular locality, especially when the firms are concentrated geographically, as is often the case in the metal finishing industry.

The statistics reported in Chapter I indicate that cut-throat pricing ranks high among the major problems facing individual firms and the industry as a whole. Respondents to the questionnaire and those approached in the interviews felt this problem exists primarily because (1) a substantial number of managers lack sufficient knowledge of the cost structure of their firm in particular, and market conditions in general, to price effectively, and (2) excess capacity exists, particularly in certain geographical areas, brought about by ease of entry into the industry and some captive shop activity, which encourages depressed pricing. The situation is serious for the industry in general; it is devastating for some firms in particular locations. The following statements summarize respondents' answers to the questionnaire and those received during interviews regarding pricing practices they have observed.

1. In many cases plant managers do not know their costs, resulting in cut-throat pricing by many firms.

2. We cannot bid a sufficient price on many of our jobs because of cut-throat pricing.
3. Many customers will get five quotations for a job. Invariably at least one quotation is too low because that manager made errors in estimating the job or doesn't understand his costs, and he gets the work.
4. We have a lack of work and resulting low profits due to improper pricing by competitors.

In fact, 95 percent of those responding to the questionnaire indicated they had lost business to a competitor because their prices were "too high." One reason for this is cut-throat competition. Another is that many managers do not estimate costs well, and overbid on some jobs. These reasons are reinforced by statements of 92 percent of the respondents—they had received jobs on which they bid, only to discover they had bid too low and lost money.

Consequently, the problem is severe as viewed from the vantage point of managers of metal finishing firms. Our field survey supported these statements, indicating that a significant number of price setters in the industry are not knowledgeable enough about the cost structure of their firms, or the market for metal finishing in general, to be effective in pricing their output. Many firms, then, would find it advantageous to reconsider their pricing policies and make adjustments and improvements where necessary.

Why Is Price Setting a Problem at All?

Why is pricing a problem at all? Doesn't the market establish prices more or less independently of an individual firm? Doesn't a firm, therefore, have to price at exactly the figure established by the market? The answers would be yes only in a setting of pure competition, since only then would buyers and sellers have full and complete knowledge of the market. In this situation, no buyer or seller would be large enough to exert substantial influence on the market, and there would be relatively easy entry into the market. The forces of supply and demand would set prices more or less automatically under such conditions. Buyers would know the exact value of the product or service to them and the prices of all competing sellers. Sellers would not only know their costs completely but also the costs and selling prices of all their competitors. One firm could not establish prices above the competitive price since customers would go to another firm to buy the product or service, and prices would fall to a new, lower equilibrium point. If demand at the lower price exceeded the firm's capacity to produce, prices would again rise until demand dropped to the point where sufficient capacity existed to satisfy it.

If an industry tried to sell at prices above the competitive price, the market would not accept the product or service, and prices would have to be reduced if substantial idle capacity was to be avoided. If prices in an industry were too low, demand would exceed the capacity to produce, and prices would have to rise to equate supply and demand, at least in the short run. Of course, over the long run, capacity can be increased or reduced for individual firms as well as for an entire industry. In pure competition, price is a perfect regulator.

Such assumptions underlying pure competition often do not hold true in the real world, certainly not for the metal finishing industry. But if we compare what is true in the metal finishing industry with these standards of pure competition, we can get a better feel of why cut-throat pricing exists. The metal finishing industry is made up of small firms and, as far as we could ascertain, no one of them is large enough to influence the market in any substantial way. Metal finishing firms are sandwiched between suppliers of raw materials and buyers of their products and services — both of whom are large enough to influence the market. There is relatively easy entry into the market. There are pockets of excess capacity in certain locations of major finishing activity. Many managers do not know their own costs, let alone those of their competitors. Putting all of these together, it is not surprising that we found many instances of pricing below what was necessary to make a rate of return on investment commensurate with the relatively high risks associated with metal finishing. We also found a significant number of cases, even worse, where products and services were priced near the out-of-pocket costs.

In any case, since the assumptions of pure competition are not valid in a number of major respects, pricing is a critical task which must be performed effectively if finishing firm managers are not to act irrationally. In the long run, irrational pricing behavior by an industry or by a firm within that industry is not good for the customer or the industry. Each metal finishing firm must take an active role in establishing rational prices for its products and services. The purpose of this chapter is to provide some broad guidelines for reaching that objective. Chapter VII will get into the specifics of pricing by firms in the metal finishing industry. But we hope the reader will not write off this chapter as too theoretical to be useful. The approaches and guidelines established in this chapter must be understood reasonably well (1) so that the reader is convinced of the importance of rational pricing and, therefore, that corrective action is needed by many firms in the metal finishing industry, and (2) to act as a foundation for understanding and accepting the recommendations in the following chapter.

A General Philosophy of Pricing

A number of fundamental ideas regarding a general philosophy of pricing must be understood and accepted before a firm attempts to establish prices for specific products and services.

Establishing the Primary Goal of the Firm

The size of profits—in fact the very idea of profit as a goal—has been attacked in some quarters for many years, particularly during the last decade or two. This is an outgrowth of the dislike and distaste on the part of certain individuals and groups for the business sector in particular and the free enterprise system in general. We choose not to debate those issues here. Suffice it to say we do not agree with such assertions.

To us, the basic objective of a firm is to maximize long-range profits within acceptable moral and required legal boundaries. Certainly, to a firm in financial difficulty, survival may be of greater immediate concern than profits. And in the short-run a firm may be more concerned with pricing to enhance the company's image or to increase its share of the market as long as profits are satisfactory, though below the maximum possible. Careful consideration of these objectives, however, indicates that they are reconcilable with, and basic to, long-range profit maximization. If a firm can't survive, it can't have profits. A good image certainly results in higher profits than a bad one. And so on. In any case, a firm, and indeed an industry, must achieve a long-run rate of return commensurate with the risks inherent in its operations—or fail. A sufficient rate of return is more likely to be attained if long-range profit maximization is accepted as the goal.

Some Fallacious Assumptions and Attitudes in Pricing

Frequently things assumed as true are actually over-simplifications of real-world circumstances. Some believe that a desired profit figure is simply added to total costs incurred in producing a product or service in order to establish its selling price. In answer to our question "Do you establish prices to quote basically by estimating total costs and adding a profit percentage?" 61 percent of the respondents to the questionnaire answered yes. Such an approach to pricing is effective only in a monopolistic environment, if at all; certainly the environment of metal finishing is not monopolistic. Even in a monopoly situation, however, pricing is not that simple because of complex joint cost problems. Our field interviews for example, indicated that many managers, even those having a reasonably good understanding of their overall cost structure, have considerable difficulty in allocating joint costs such as transportation costs, various selling and administrative costs, and certain fixed manufacturing costs for price-setting purposes. Also, this attitude of adding on a desired profit to total costs completely overlooks the importance of cost control—the standards and procedures necessary to utilize cost factors (materials, labor, overhead) in the most efficient, effective manner. Any firm which attempts to build the costs of inefficient utilization of factors into its pricing structure is not likely to stay in existence for a very long period of time, particularly if it operates in a reasonably competitive environment. In sum, costs do not control prices charged, although they do frequently influence them.

Others seem to believe that prices have complete control over costs, i.e., that sellers take the established price, reduce it by a necessary profit margin, and this establishes the costs of production (amounts the sellers will pay for materials, labor, and capital.) Obviously, this is not possible to any large extent in the metal finishing industry. First, as discussed earlier, material costs seem to be increased frequently and almost at will by suppliers, and metal finishing firms have no substantive influence in changing those prices. Second, labor costs cannot be changed significantly over short periods of time; and third, the financial condition of the firms and the risks involved in their operations establish the price they must pay for capital funds. It is true that pressure from established selling prices might, and perhaps should, cause firms to devote considerable energy toward using cost factors in the most efficient and effective way (and such efforts are long overdue). We found some managers, for example, who said they searched for cheaper material substitutes, plated less thickly, and took other steps to reduce cost factors as long as minimum quality standards were met. Nonetheless, prices paid for cost factors can only be established on a "selling price less profit margin" basis in a purely competitive situation, and we have established previously that the present environment of metal finishing is not one of pure competition. In sum, although prices do influence costs, they do not regulate them.

Furthermore, concentrating strictly on costs or selling prices overlooks the effect of volume or capacity. Capacity available, both at the firm and total industry levels, influences costs and selling prices and in turn is influenced by them. Cost increases brought about by decreasing volume or higher overhead often cannot be passed on to the customer. Selling price decreases, on the other hand, may increase production activity, thereby decreasing the average cost per unit of work done and increasing profits.

Consequently, often there is no single price that can be established. There is, instead, a range of prices within which a sale can be made. Certain costs (out-of-pocket) do establish a minimum below which the product or service should not be sold. The upper limit, in given circumstances, might be established by customer demands, such as short lead times, and the absence of competitive pressure (e.g., only one firm in a given area performing a given type of finishing work). When competition is severe, as is true when excess capacity exists in an industry or a geographical or a production-type segment of an industry, prices may be pushed down to a level nearly equal to out-of-pocket costs, presently a much too common situation in the metal finishing industry. If demand is high for a product or service having no substitutes, price may rise far above the costs incurred in producing and selling the product or service. When excess capacity does exist in certain locations, and when entry into the industry is relatively easy, as is true for metal finishing, it becomes even more critical for managers to understand the long-range consequences of their pricing policies, both from the vantage point of their individual firms and that of the industry in total.

A discussion of fallacious assumptions and attitudes in pricing would not be complete without consideration of the responsibilities of the customers of metal finishing firms. All buyers like bargain and distress prices, and many enjoy taking an active part in negotiating and determining prices. Certainly there is nothing wrong with this in general, whether the buyers involved are customers of the automotive companies, the home appliance industry, the metal finishing industry, or any other. A short period of consideration should make it clear, however, that no firm or industry can survive long by making such prices continuously available. Nor would it be to the customer's long-range advantage for it to do so. Selling constantly at distress prices certainly would result, at the least, in a rate of return to the seller less than commensurate with the risks inherent in his operations; at worst, the prices would not even cover his costs. In either case, the results would be the same—a gradual liquidation of the business. This liquidation process might take some time, but inevitably the supply of the product or service would be cut off because capital would be transferred into other, more profitable ventures. When the service or product is to become a component of a customer's product—the case in metal finishing—the customer will be forced either to drop the manufacture of the particular product or to set up a captive shop to perform that particular service or operation. Whichever course he takes, his profits may be adversely affected. The customer, in his zeal to force reductions in price in the short-run, may be seriously endangering the long-run supply of a service or product needed to maintain his operations. Certainly we have seen that happen in the metal finishing industry, with the result being a substantial increase in captive shops from the late 1950s through the late 1960s.

Principles for a Sound Pricing Policy

In establishing selling prices, future prices of the cost factors should be anticipated or estimated. Past and even present costs are not usually the best cost figures to use in price determinations. This is particularly true in periods of rapidly changing prices, or where prices are quoted on a job to be performed some months into the future. In spite of the fact that most materials used in finishing operations have fluctuated widely (usually in an upward trend) over the past few years and replacement costs for plant and equipment have increased dramatically, we found few managers who were estimating future costs and using them in pricing decisions. When anticipated replacement values for material, labor, and investments in plant, property, and equipment are not used, capital investment will erode; higher prices and wider profit margins are needed to generate the funds from operations needed to replace these items.

The one responsible for setting prices must have reasonable knowledge of the cost characteristics of his firm and must use the appropriate concepts of cost when making pricing decisions. Without such knowledge and use, not only are incorrect pricing decisions likely to be made but the price maker will not be aware of the consequences of his decisions, once made. Of the questionnaire respondents, 75 percent said they made a clear distinction between fixed and variable costs in price setting, and 80 percent said they used some form of standard costs for pricing purposes. Although these statistics are considerably better than those in 1967, we remain somewhat skeptical that things are as good for the finishing industry in general as these figures might imply.

First, 61 percent said they simply add a profit percentage to total actual costs to establish prices, and this is inconsistent with the other two answers. Second, although we made no effort to determine the size of respondent firms in order to preserve confidentiality, we suspect the larger firms with the better accounting systems made up most of the respondents. Third, our field survey indicated that many managers consider "standard" to mean some fixed amount per barrel (such as $10 or whatever), often based on actual costs of some months or years earlier. This, of course, is a far cry from the meaning of "standard" used by accountants. Fourth, only 66 percent consider the volume level at which they are then working when quoting prices, implying that not as much weight is given to fixed costs in price setting as previously indicated in the other answers of the respondents. Finally, our field survey clearly indicated, though only a small sample, that proper weight is not given to fixed and variable costs and to standard costs in price determination decisions.

In any case, standard costs, both fixed and variable, should be utilized more in price determinations than presently is true, assuming, of course, the standards established are reasonable. Costs above reasonable standards cannot be passed on to customers in higher prices in the competitive market of the finishing industry. Such differences must be eliminated through appropriate cost control techniques. Thus, in some cases, good cost control procedures would make it possible to decrease prices and increase competition.

The price maker also must keep abreast of market conditions generally and conditions in his industry in particular, especially mid-to long-range forecasts regarding both. He should strive to gain whatever knowledge he can of customer desires, of the utility or value of his products and services to the customer under varying conditions and circumstances, and of his competitors' estimated and anticipated costs. Perfection in this regard is not possible, of course, but the best individual at making pricing decisions is the best informed one. We found too often that prices are being established in a virtual vacuum—a critical problem in the metal finishing industry. Under such circumstances, price makers have little knowledge or understanding of the consequences of their pricing decisions on their own firms or on the industry generally.

To illustrate what we mean, we can look at metal finishing as a service function where total volume or demand for the industry is fixed by the needs of the various industries served. Changes in prices within reasonable bounds for finishing work, therefore, are not likely to affect total volume of the finishing industry. Certainly this is true for price reductions, since volume would be increased for the independents in total only if captive shops were closed—not a likely occurrence. It is also true for price increases, unless they become so high that captive shops are established. Consequently, even cut-throat pricing is not likely to increase total demand for finishing services. More likely it will result only in switching demand back and forth between individual firms (especially in those areas having excess capacity), with all the short-run benefits accruing to the advantage of the customers. Everyone is likely to be hurt in the long-run for reasons previously enumerated.

A final basic principle is this: costs should be broken down for each type of finishing operation wherever possible and used as guides in pricing. When this is not done, (and we found it frequently is not,) losses on some types of finishing work may go unnoticed because they are being covered by high margins on others. A breakdown of costs helps to determine which type of finishing should be reduced or dropped, and which should be promoted and expanded. It helps to direct marketing efforts and to determine which jobs or customers should be dropped, or not accepted in the first place, because of low profit margins.

Bringing the General Principles into Clearer Focus

Long-range profit maximization has been emphasized as a goal in this monograph. Certainly there are difficulties in attaining that goal because of many demand and supply uncertainties in the marketplace.

Establishing a target rate of return on total assets employed, or on the owner's investment in the business, is a good starting point on which to base prices—and one apparently seldom used in the metal finishing industry.

Establishing a target rate of return is not a simple process. A number of considerations should be taken into account, such as the return that might be earned on capital invested in ventures of comparable risk, a weighted cost of capital, and published normal or average rates of return for firms in the industry. We emphasize that this is a starting point only. The target rates of return and the target prices based on them are not meant to imply that such returns can be earned on every sale, or even than an average rate of return can be earned over the short run on all sales. They do specify a goal which must be attained (or nearly so) over the long run if the firm is to stay in business at a reasonable profit level. Target rates of return of 20 to 40 percent before taxes on total assets employed are fairly common in American industry under normal economic circumstances. (Certainly these rates fell during the depressed 1974-75 period.)

To see how using target rates of return works in general, assume the balance sheet data which follow.

METAL FINSHING SHOP BALANCE SHEET
As of December 31, 19—

Cash	$ 15,000	Short-term Payables	$ 45,000
Receivables	22,000	Long-term Payables	25,000
Inventories	43,000	Capital Stock &	
Plant & Equipment	300,000	Retained Earnings	
		(Owner's investment)	310,000
	$380,000		$380,000

Assuming that a 40 percent rate of return before taxes on the owner's investment is a reasonable target, then $124,000 (40% × $310,000) is the targeted profit before taxes. Assuming fixed costs are $150,000 per year and the contribution margin percentage (net sales less variable costs divided by net sales) is 40 percent, then net sales would have to be $685,000 in order to provide a profit before taxes of $124,000: sales = (sales - 60% sales) + $150,000 + $124,000. In projected income statement format, these resulting figures would appear as follows:

Net sales	$685,000	100%
Variable (incremental) costs	411,000	60%
Contribution margin	$274,000	40%
Fixed costs	150,000	22%
Profit before taxes	$124,000	18%

A number of observations are possible from these data. An 18 percent margin of profit before taxes on net sales equals a markup of 22 percent ($124,000 divided by $561,000) on total costs. Investment turnover is equal to 2.2 times per year ($685,000 of net sales divided by $310,000 of owner's investment). The required profit percentage on sales can be quickly determined by dividing return on investment by the investment turnover (40% divided by 2.2 = 18%).

Although we develop specific examples of pricing in the next chapter to reflect the application of these principles, let us complete the example here by assuming the company has only one customer, finishes only one component for that customer, and the anticipated volume for the coming year is 1,400,000 units. Under these assumptions, the metal finisher would have to charge 48.9 cents per unit ($685,000 target net sales divided by 1,400,000 units) in order to attain the targeted rate of return on owner's investment.

Does it necessarily follow, however, that he can or should charge 48.9 cents per unit plated? Not necessarily. The finisher should consider competition in the area, the demand for the product of his customer for which the component is a part, and the other factors previously discussed. If the market will bear a higher price than 48.9 cents per unit, it should be charged. Or the market circumstances may be such that a lower price must be charged. In either case, the consequences of a decision can be readily observed with this approach. For example, if price is decreased by one penny per unit, profit before taxes will decrease $14,000 (1,400,000 × $.01). This will decrease rate of return on investment to 35.5 percent ($124,000 minus $14,000 divided by $310,000) — down 4.5 percent below the target rate of 40 percent.

Of course, when more than one customer is served and/or more than one type of finishing work is performed, the problem becomes more complex. As indicated, we will give pricing examples under more complex circumstances in the following chapter. Suffice it to say here that if prices could be set at an 18 percent margin on sales (profit before taxes divided by net sales) for each job or service performed, the firm would end the year by achieving its 40 percent target rate of return before taxes on owner's investment, if we assume the figures given in the example. This would be possible only in rare instances, of course, since many factors other than costs influence prices, as we have discussed. To the degree possible, the objective is to price so that the profits on all jobs combined will average out to at least the target rate of return.

Costs should be used to indicate ranges within which prices may be set with known profit results and to provide a basis for calculating target prices. As previous discussions indicate, this requires analyzing costs to reflect their fixed and variable natures so that incremental costs and incremental profit margins can be measured. These are the data of significance to the price maker for individual jobs and services when more volume is needed; they help to determine where marketing and sales efforts can be expended most beneficially, to determine which customers and/or services should be dropped, and to determine how to react to severe competition. Close figuring is essential, and it cannot be accomplished effectively by use of an arbitrary formula nor by "seat of the pants" reckoning, even for the smallest firm.

In those instances where a shop performs a number of different kinds of metal finishing work, and where competition for the output of the different operations varies substantially, rates of returns are better applied to incremental (variable) costs rather than to total costs. An important advantage of this approach is that it avoids distortions possible when total costs are used, requiring arbitrary allocations of many joint or common costs. Depreciation on the building housing the operations, trucking costs, and many of the selling and administrative costs are examples of common or joint costs where allocation to particular jobs would be arbitrary at best, since no direct causal relationship can be shown between them and the jobs performed. Since allocation is arbitrary, more or less may be allocated to one job than is warranted, which might result in losing a job because it is overpriced or losing profits because it is underpriced. Better, in such circumstances, to use an incremental cost approach.

For example, using data in the preceding example—net sales of $685,000 and variable (incremental) costs of $411,000—the contribution margin must be $274,000 in order to cover fixed costs of $150,000 and to provide a profit before taxes of $124,000 (40 percent return on owner's investment before taxes). Since incremental costs are expected to be $411,000, the markup must average 66-2/3 percent. If the incremental costs of three different types of finishing are $110,000, $101,000, and $200,000, the target contribution margin would have to be $73,000, $67,333, and $133,332 respectively in order to equal a total contribution margin of $274,000 and thereby reach the target profit of $124,000 before taxes. This assumes that only costs are considered. Taking into account supply and demand factors in the market, the price maker might determine the target contribution margin for each type of finishing respectively is $68,500, $85,400, and $120,100. This would establish target markups on incremental costs approximately 62 percent, 85 percent, and 60 percent respectively for the three types of operations. In determining quoted prices, the target price and its cost components should be kept in mind.

Costs are used to indicate whether management will ever accept a job at a given price, under what conditions certain jobs will be accepted at certain prices, and whether or not the goal or objective in pricing (maximization of long-range profits) is being attained. Incremental costs represent the minimum price that should be charged for a job. (We found instances where prices were quoted lower than incremental costs because the price quoter did not know his costs or was using them incorrectly). Price should be lowered to the incremental cost level only under certain compelling circumstances. The results of a decision to charge this minimum should also be clearly understood—which frequently is not the case, or the price setter would not be quoting that price in the first place.

When a product or service is priced at the incremental cost level, presumably there would be no effect on profits. There also is likely not to be any immediate effect on the working capital of the firm. The long-range effects, however, could be serious. First, the investment in plant, machinery, equipment, and all other items previously acquired is being eroded. If such pricing policies were continued for an extended period of time, the business would eventually liquidate, since the cash inflow from sales would be inadequate to replace plant and equipment as needed. Second, although prices are easy to lower, they are often difficult to raise. Once prices have been established at an incremental cost level, customers may expect them to continue or may vigorously oppose increases. Third, cutting prices to this level would force competitors to do likewise and would precipitate a downward spiral which might be very difficult to counteract.

Yet there are cases where a firm might logically sell at a price approximating incremental costs. When an entire industry is in a serious economic slump and incremental cost is as high as the market is willing to go to acquire the product or service, firms might accept business at this price level in order to keep their work force, management personnel, and supply lines intact. Equipment and machinery often continue to depreciate when not in use (and some types depreciate more when not in use), so the firm might be better off operating under these conditions than closing down.

Even if a firm never sells at this low level, there is another important reason for having information on the incremental costs for each type of operation. Any price above this level contributes to the fixed overhead. Once total fixed overhead is covered, each dollar of contribution margin (sales less incremental costs) increases profit before taxes by a dollar. Knowledge of profit contribution is necessary if the firm is to determine the mix of metal finishing operations that will lead to maximum profits. Management is then able to emphasize the type of finishing work which makes the greatest contribution toward profit, and to reduce or drop certain types of work (where feasible) when the market is unable or unwilling to pay a price that makes a reasonable contribution to profits.

The relative weights of material, labor, factory overhead, distribution, and selling expenses should be considered, because they vary with each type of operation and also over time. Furthermore, unless a more nearly complete cost per type of job or operation is known, it is impossible to know if one operation is subsidizing another or why the price of a particular operation is not competitive. Full cost per type of operation is important to permit management to see the consequences of a pricing decision. This information allows management to make a more informed decision about accepting or rejecting business at a given price.

Here, again, certain conditions may influence management to set a price between incremental costs and full costs, though this is not an ideal practice. It there is idle plant capacity and the job is a one-time order requiring only a short time to complete, management might well decide to take it at a price higher than incremental costs but less than full costs. Every dollar above incremental costs contributes to fixed overhead (and hence increases profits or reduces loss) and is a gain over what would be obtained if the capacity were allowed to remain idle.

A word of caution is necessary. Management should not, as a general rule, accept jobs on this basis which are going to be repeat jobs, nor should they accept on this basis jobs which will extend over a considerable period of time. To accept repeat jobs (meaning jobs that may be handled repeatedly into the foreseeable future) at a price between incremental and full costs is to establish a price that the customer will continue to expect in the future. Jobs of long duration accepted on this basis reduce the flexibility of the firm and make it difficult, if not impossible, to take advantage of high-priced jobs when they are available.

Of course, the objective of management is to sell at a price which will result in long-range maximization of profit. The target rate, generally, should be looked upon as the minimum acceptable long-range rate. Since this is the goal, target price information similar to that which we have been discussing should be constantly in front of the price maker as an aid in reaching that goal. A target price should not prevent his accepting jobs where the market has set a higher price, of course. But this information, along with information of full costs and incremental costs, makes price setting (or the decision to accept business at a price set by the market) more scientific, and it makes the results of the decision clear.

When the economy is running at a high level, the management of a successful firm is tempted to play down or be unconcerned with the pricing problem. Yet the concern for proper use of pricing techniques should be almost as great under these conditions as it is for the management of an unprofitable firm. When a firm is operating at near full capacity levels, it must exercise great care to ensure a mix of metal finishing operations that will maximize profits. The jobs which are then accepted should be those entailing operations that are likely to yield the highest contribution to profit. Otherwise plant capacity will be used up on low-yield jobs, and even though the operations might still be profitable, they would not be as profitable as they ought to be. In other words, a ranking scheme needs to be devised in order to attain maximization of profits.

Moreover, there are bound to be low levels of economic activitiy during which companies will operate at considerably less than full capacity. (A number of metal finishing firms are in this position at present.) Cost information and target prices as described above will help such firms to hold profits at the highest possible level—or losses to the minimum possible level. During slack economic periods, the question is often not which firms make the greatest profits, but rather which firms suffer the least losses. It is the latter companies that survive the slump and reestablish their profit pattern when conditions change.

A Few Words of Caution

Some words of caution are in order on a number of points. Throughout this monograph we have emphasized the necessity for remaining within legal and moral boundaries when developing pricing policies, merging, or in any other action that management is contemplating. We have also indicated that our suggestions usually are general and broad in nature because this was our primary charge and also because the wide divergences in the metal finishing industry made this necessary. Our recommendations are not intended in any sense, nor should they be so interpreted, to suggest or imply anything illegal or immoral. Metal finishers, like all businessmen, must seek advice from experts, particularly legal counsel, in developing and implementing specific techniques and procedures. Unfortunately, we live in a "legal-suit happy" society, and this must be kept in mind when actions are taken.

Problems other than potential legal problems exist. For example, in making calculations of target rates of return, investment should be measured in terms of current or replacement values. Otherwise capital will be eroded during periods of rising prices, and quoted prices will likely be too high during falling prices. We recognize that adjusting historical costs to a current value basis can be time consuming and costly. At the minimum, the price setter should consider such current values in terms of whether they are higher or lower than historical costs and accordingly make a rough adjustment to the target rate of return.

Finally, the target profit before taxes figure should be set after a reasonable deduction for "wages earned" by the owner, whether removed from the firm or not. The target rate is applicable to the capital investment of the owner and should not include any return to him for services performed for the firm. Frequently, the amounts withdrawn from the firm by the owner are not a good measurement of the values of his services to the firm. The amount deducted should be the best estimate of what the owner could command by way of wages or salary if he performed the services for another firm as an employee.

In this chapter we have been discussing what "should" be done if long-range profit maximization is to result. Obviously, an owner can accept less if he is willing. But he ought to be fully aware of the consequences to himself, his firm, and the metal finishing industry in general of the decisions he makes.

Summary and a Few Conclusions

Although some firms are doing some of the things recommended here, there is a need for improved practices. A large number of managers do not make a distinction between fixed and variable expenses for pricing purposes, and many did not appear to be well-informed about the proper use of such information. Most consider historical costs only in setting prices. Few seemed to make any estimates for material, labor, and overhead costs for the coming year for use in establishing prices. Many had little idea of the success or failure of their pricing decision until they received monthly or quarterly profit and loss statements. Even then, many were using information which did not clearly disclose the consequences of the decisions made.

The metal finishing industry can benefit from the use of more scientific methods of establishing prices. This could lead to the elimination, or at least reduction, of cut-throat pricing; it might lead to a greater degree of specialization where warranted and to larger long-range profits. We feel that achievement of these goals can help to improve the industry as a whole as well as to increase effective competition in society.

VII

PRICING FOR PROFIT

Inferior job costing and pricing is a critical industry problem that was cited by several managers in this survey. Cut-throat pricing often is a consequence. One of the basic causes of the problem is that few firms have effective cost accounting systems which accurately collect and record detailed costs of jobs. Without such job cost histories, managers frequently do not understand their current cost picture and thus are at a disadvantage when estimating costs on all but the most common jobs. The impact and severity of these problems were articulated by one plant manager in the following statement:

> Pricing and profitability within our industry is a most significant problem. Most owners in the industry have either gotten into that position by family relationships or have come up through the ranks. In general, there is limited sophistication within the industry as far as management prowess is concerned. Thus, many members of management are limited in the knowledge and techniques required in knowing which jobs they are and are not losing money on.

Another manager alluded to the inferior cost estimating and pricing methods by citing the following experience.

> At an industry seminar on pricing methods, a sample part requiring chrome plating was passed among the several managers present. Each was asked to estimate what the costs would be for his company to plate the part. After the part had been inspected and the costs estimated, each manager submitted his written cost. Upon opening the estimates, they were found to vary from $1.15 to $6.25, a range of nearly five-fold.

Although some variation can be expected, this example makes clear the significant difficulty many managers have in estimating costs. Without such information, pricing becomes an exercise in futility at best; at worst, it is a damaging exercise both to the firm and to the industry generally.

In Chapter VI, we discussed many of the principles and general procedures underlying a sound approach to pricing. Though general, we emphasize again that the information presented there is not to be ignored if managers are to improve their pricing abilities. Based on that foundation, in this chapter we will consider the specific problems managers in the metal finishing industry have in pricing effectively, and we will make suggestions for improvements.

General Cost Estimating and Pricing Philosophies Presently Being Used

One of the difficulties in generating reliable cost information is that a shop may (and frequently does) finish parts of varying sizes and shapes with several different finishes requiring substantial process variations. Under such conditions, it is difficult to ascertain and keep current the actual costs associated with each stage in every process. The many process variables make it extremely difficult to set standards applicable for all part shapes, sizes and finishes. In the absence of reasonably complete, accurate, and discriminating cost data, the basis of job estimates frequently is a "not-so-educated guess." Any similarity between that and a correct estimate is likely to be happenstance at best.

Designing and implementing an economic cost accounting system is a problem. Cost estimating efforts alone can consume potential profits on small volume jobs of, say, 15 to 100 dollars. The system, therefore, must be designed to provide reasonably accurate information for large volume jobs and acceptably rough approximations for low volume jobs. Many systems we observed did neither, leaving managers with little if any knowledge as to which jobs were profitable and which ones were not.

Before discussing a more scientific approach to pricing, a short description of what we observed some companies doing might be helpful as a foundation for indicating strengths and weaknesses in approaches we have observed.

Labor Based Costing

Labor based costing is frequently used in manual or non-mechanized finishing companies where labor constitutes the largest proportion of total costs—due to the time and effort necessary to prepare, load, and unload the parts, and to convey them manually through the various cleaning, finishing, rinsing, and drying stages. Job costs are determined by estimating the number of direct labor hours a job will take and multiplying this times the hourly labor rate plus fringe benefits. A predetermined overhead rate (including all company costs except direct labor) is then multiplied times the direct labor hours consumed on the job and added to labor costs (including fringe benefits), in order to arrive at total costs for the job. A profit element is then added to total costs to arrive at a quoted selling price. Note that materials are not estimated as a separate component, nor are overhead items broken down between those directly traceable to a particular line (such as depreciation on the equipment) and those of a general nature not directly traceable to any particular line or operations (such as depreciation on the building, or selling expenses).

Such a procedure has serious shortcomings except in very specific circumstances. For example, if there is only one finishing line, if materials are an insignificant part of total costs, if overhead costs are an insignificant part of total costs, and if labor costs are the dominant proportion of the total, such a procedure may be acceptable. Such conditions, in total, are not likely to exist in even the smallest of firms. Even in small nonmechanized firms, overhead costs such as depreciation, repairs and maintenance, and general and administrative costs are likely to be significant enough that, if they are estimated or allocated incorrectly, profits might be eliminated on many jobs through underestimates or many jobs lost through overestimates of total costs. In addition, although material costs do vary substantially depending on the type of finishing being done, it is fairly rare for materials to be so insignificant that bad estimates won't affect profits substantially. For example, if we assume this method of cost estimating is used to estimate the cost of plating two identical parts, one with a heavy finish and the other with considerably less, the same total cost would be estimated for both jobs, even though the difference in actual materials used might be substantial. Such differentials increase, of course, as the price of plating metals increases (such as when gold or silver are used).

Even when the basic price per unit for the materials is low, a significant variation in usage might cause the manager to price some jobs too low and reduce profits, or price some too high and lose jobs to competitors with better cost-estimating procedures. Also, usually it is not the best course to allocate nontraceable (indirect) overhead (e.g., general and administrative expenses) to jobs on the same basis that traceable (direct) overhead is charged (depreciation of equipment on a particular line).

Finally, when these procedures are utilized, it becomes very easy to use an overhead rate for a number of years without serious question as to its current validity. We found a number of examples where the overhead rate used had been derived a number of years earlier and had not been updated on a current cost basis. Even if such costs are immaterial in any given year as a proportion of total costs, inflation can cause a substantial difference between current costs of such factors and those utilized in developing an overhead rate which is several years old.

Because direct labor in manual finishing companies averages about a third of total costs, we found some companies take a "short-cut" to a price quotation by multiplying the direct labor estimate by three, and adding a profit factor. This, of course, is even a much rougher estimate than that previously described. In addition to the arguments against such a procedure already presented, we found that materials as a proportion of labor expenses varied from one-third to three-fourths in labor-intensive companies in the NAMF cost surveys. The extent to which labor and material ratios vary throughout the industry is reflected in the following table:

TABLE 4
RELATIONSHIPS OF LABOR AND MATERIAL COSTS
PARTICIPANTS IN 1970 NAMF SURVEY

	Expenses as a Percent of Sales		
Classification of Company	Direct Labor	Materials	Materials as a Percent of Direct Labor
Manual - Nonmechanized	29	16	55
Manual - Barrel	20	22	110
Automated	25	22	88
Engineering*	26	17	65

*Includes hard chrome, electroless nickel, hard anodize.

Basing job cost estimates on historical averages and multiples of labor cost, when it is the major variable expense, has led to serious pricing errors, especially when excessive process requirements were encountered. One flaw of the "averaging" concept that was observed repeatedly was the failure to take into account "unusual" costs for such requirements as pickling, isolated buffing, special tests of plated material, baking to remove embrittlement, and extreme transportation distances of certain orders, to name a few. For pricing purposes, companies often fail to add such costs to the "average" price derived as a multiple of direct labor. Problems with this philosophy also occurred during 1973–74 when material prices fluctuated so drastically. As a consequence, extra-cost jobs were underpriced, while the nonspecial orders were overpriced.

One of the most serious consequences of such unscientific procedures is that they become so routinized that basic principles essential to effective pricing, as discussed in Chapter IV, are ignored or overlooked. Prices are assumed to depend exclusively on costs, when the material in Chapter VI indicates clearly this is not always—or even generally—the case. The practical consequences of such flaws vary, but they can be significant in certain instances. More than one company has found itself with falling profits as it obtained too many underpriced jobs and lost orders it overpriced.

Equipment Based Costing

This procedure predominates in companies doing production finishing with mechanized or automated equipment, and sometimes in nonmechanized finishing characterized by equipment-dominated expenses, such as in hard chrome plating. Some hard chrome jobs may need to be immersed in plating tanks for a day or more, and cost estimates for these jobs should include an appropriate share of tank and equipment related charges. When companies use equipment-based costing, equipment related expenses often comprise the major cost component, with direct labor costs falling commonly in the range of 12 to 18 percent of the total.

When equipment or tank based costing is followed, usually all costs associated with operating the equipment are aggregated for some past period, frequently a year, and are then divided by the number of operational hours over that same period. The result is an hourly operational cost. Production per hour can be divided into the hourly operational cost to ascertain the indicated cost per pound or piece. Commonly the rate includes plating metals, cleaning, rinsing, other chemicals, gas, electricity, water, supervision, equipment amortization, and the like. If production labor costs tend to remain steady from job to job, say, at three, four, or five workers on the line, these costs usually are also included in the aggregated equipment costs. If labor varies significantly, the costs are estimated separately and accounted for as direct labor.

Labor to load or unload the parts frequently is calculated separately, and these costs are then added to the equipment operational costs. Other costs may also be calculated and added for special tooling, such as racks and conforming anodes. Subsequent operations such as buffing and polishing are other examples of elements costed separately.

The content and relative magnitude of equipment operational expenses for one survey company's barrel plating machine is shown in Table 5.

TABLE 5
EXAMPLE OF EXPENSE ACCOUNTING FOR BARREL MACHINE

	Annual Amount		Percent of Total
Fixed Costs:			
Amortization of equipment cost	$ 5,925		9%
Scheduled maintenance	500		1%
Total Fixed Costs		$ 6,425	10%
Variable Costs:			
Direct Labor	$10,987		17
Supervision	7,387		11
Electricity	3,744		6
Water	3,639		6
Materials	10,429		16
Anodes	7,552		11
Steam	5,553		8
Insurance	527		1
Unscheduled Repairs	9,687		14
Total Variable Costs		59,505	
TOTAL ANNUAL COSTS		$65,930	100%

Notice that this system includes labor and material costs in building up an hourly equipment rate. Note also that the expenses are broken down between fixed and variable, a necessary beginning step if effective pricing is to result. Because this machine was used on a two-shift basis (4,160 hours per year), the hourly operational rate was $15.85: total annual cost ($65,930) divided by 4,160. When a job is bid, the number of parts the machine can plate per hour is calculated, taking into account such things as the part shape, thickness, type of metal to be plated, and equipment amperage. The costs are then stated on a per unit basis or, more commonly, in terms of 100 units. In our example, a processing rate of 500 pieces per hour resulted in a cost per unit of 3 cents ($15.85/500), $3 per hundred, or $30 per thousand.

There are some good features in the system described. An attempt was made to separate costs into the variable and fixed components (though, as indicated later, they could have been used more effectively). Also, there was some effort made to charge special costs, such as special racks and conforming anodes, to the particular job creating them. This is an attempt to follow a basic principle in cost estimating; i.e., *generally, all costs directly traceable to a given job should be charged to that job and no other.* In this sense, the word "direct" means the practical, obvious tracing of cost as incurred to the job without any intervening allocations of that cost being necessary.

But there are a number of bad features in the procedures followed. First, although a careful consideration as to the manner in which costs behave for this firm in relation to volume changes would be necessary to arrive at specific conclusions, it appears that some of the costs were misclassified between variable and fixed. For example, over given ranges of volume, supervision is usually fixed, not variable. Insurance is usually fixed. Some of the other costs (such as steam) are probably mixed costs and need to be separated into fixed and variable components.

Second, materials are a significant component of total costs in the example (15.8 percent), and will not necessarily vary per job in the same proportion that labor hours vary. Some firms, recognizing this, compute materials separately per job. Third, past costs are used by this firm in pricing. Anticipated (projected) future costs ought to be used. Past costs can act as a springboard or foundation for forecasting, but estimated future costs ought to be used. Fourth, past periods' volume is used for determining cost per direct labor hour. Again, the past is useful in predicting the future, but anticipated volume for the coming year should be used.

Finally, fixed and variable costs should be stated independently of one another so they can be considered separately in the pricing decision. Without such an approach, as we have stated earlier, it is difficult, if not impossible, for the manager to see the effect of a pricing decision on profits of the firm.

Plated Surface Area Costing

Variations of this philosophy are used by many in the industry, but especially by companies that plate with precious metals. Some use it exclusively, while others combine it with direct labor or equipment based costing. Cost estimates for precious metal plating such as gold, silver, and platinum are largely based on material costs—or multiples thereof, relative to the square inches plated. The multiples are derived from experience and they encompass material cost, labor, overhead, and equipment expenses. Because the cost of precious metal is such a significant expense, material estimates should take special note of anticipated potential future swings in market prices, but we found this frequently was not done.

Pricing for Profit

The three cost components which need to be estimated accurately for pricing decisions, obviously, are materials, direct labor and overhead. The importance of each, relative to the others, depends upon such factors as the price of materials utilized, whether the firm is labor or capital intensive, the policies of the firm, and the activities of the manager. (If the manager spends most of his time in production activities, his salary should be allocated to specific jobs where possible, and if that isn't possible, through a departmental or finishing line overhead rate, rather than as part of the plant-wide or general overhead.) Although we found firms having varying degrees of difficulty estimating any one or all of the cost components, most of the difficulties and errors observed related to overhead cost estimating and its proper use in price quoting. Most of our time, consequently, will be spent on overhead, but some comments are appropriate for materials and direct labor.

One other important point needs to be made first. We are not covering cost control procedures and techniques in this monograph, for reasons previously stated. It is important to observe, however, that pricing should be based on reasonable standards for materials, labor, and overhead. In a competitive environment, a firm cannot charge the customer for excessive materials utilized, excessive labor used, or inefficiently utilized overhead items. Any attempt to do so will drive the customer to a more efficient competitor. We will make some suggestions along these lines; but since we are not discussing cost control features such as setting standards, analyzing variations from standards, and the like, we must assume that the cost components are being utilized reasonably efficiently. To the degree that this is not true for a given firm, the problem is one of cost control—not one of trying to recoup excessive costs through price increases.

Materials

Many firms do a reasonably good job in estimating material costs by jobs. Depending upon the type of finishing work performed, materials usually should be divided into three major elements: metal actually deposited on the part, metal "lost" during the plating cycle, and chemicals, such as those used in the cleaning, plating, and rinsing stages. The estimated cost of the metal required on the plated part can be determined by computing the volume of the metal to be used and its price per pound, or, more typically, per gram. The specific procedures that can be used are the following:

a. Determine the volume of metal required by multiplying the required metal thickness by the part area to be plated;
b. Convert volume to weight of metal to be plated on each part;
c. Multiply weight of metal by the projected metal price.

Cost estimators, particularly those working with precious metals, should check their estimates of metal to be deposited by plating a limited number of sample parts where possible—weighing the part before and after plating. The difference, obviously, is the metal deposited. Where trial weighing is not possible and the cost of the metal to be plated is expensive (e.g., gold or silver plating of sizable parts), the estimator is under extreme pressure to estimate correctly the plated area's volume and material's weight used. Too low an estimate means a loss on the job, while one that is too high may result in a price quotation that is underbid by a competitor.

Estimating "drag-out" is even more difficult. A certain amount of plating and rinsing solutions, including costly chemicals, is commonly lost as parts are processed through the plating sequences from one solution tank to the next. The impact of this so-called "drag-out" can be substantial. Drag-out differences can vary significantly between processes and also between parts of different shapes even in the same process. Parts having a cup-like configuration, for example, cause above average drag-out when the solution stalls in the cups rather than draining naturally off the parts after they have been raised above the solution tank in preparation for transfer.

Amounts of solutions lost or dragged to the next tank can be calculated by analyzing the chemical content of a new solution in subsequent stages before and after the part batch is first processed through it. The results of such a sampling analysis are valid only for that particular part and process, and a reasonably large sample of the actual parts must be available for trial processing. Nevertheless, this technique can be used to determine a generalized standard for a family of parts having similar profiles and processing requirements against which closely related parts can be compared. Drag-out can then be stated as a percentage of the metal actually deposited for various configurations, and the percentage maybe utilized in material cost estimating. This trial method is especially preferred in high-volume applications, particularly where the plating run may continue for months or years.

We have recommened that, ideally, some estimated average material costs for the coming year be utilized for cost estimating. This eliminates the necessity of changing prices frequently, which can be a problem, particularly with repeat customers. At the minimum, the price to be used should be the estimated cost to replace the materials at the time the bid is submitted, not the price actually paid for the materials used. Also, items such as drag-out, which might be stated as a percentage of material deposited, should be checked frequently. In fact, all standards established should be reexamined frequently and kept current. We found this often was not the case in practice. In any case, materials utilized on a particular job should be directly traced and charged to that job wherever possible. Materials should not be run through "overhead" and charged to jobs on the basis of direct labor hours or machine hours consumed.

Direct Labor

As is true with materials, direct labor should be estimated, traced to, and charged to individual jobs, rather than being estimated and allocated to jobs on the basis of some predetermined overhead rate. Labor, such as racking time, can vary significantly between jobs depending upon part size, configuration, and the like. Time cards should be used where the worker can enter the time spent on each job. The time multiplied by his rate per hour equals the charge to the job. A build-up of such actual experience data, where it has not already been done, can be used as a basis for estimated labor in future price setting situations.

But what should be included in the rate per hour estimates? The general answer is the basic rate per labor hour plus those fringe benefits which are variable over the anticipated volume range. For example, if a metal finishing worker does not earn up to the maximum on which social security taxes are based, then the firm's portion of the tax is variable and should be included in the hourly rate. If he does earn over the maximum amount, then the tax is fixed and should become part of the overhead to be estimated and applied as we indicate later. Federal and state unemployment taxes usually are fixed and should go through overhead since most workers will earn above the amounts on which these taxes are based. If these procedures are not followed, jobs will be overbid in the earlier part of the year and underbid in the later portion of the year.

What if some workers work overtime at a premium? If this is an ongoing routine, such premiums are fixed and should become part of the overhead, to be spread as part of the estimate on all jobs. If workers, say, work ten hours per day on the average, it cannot be argued logically that the job worked on during the last two hours caused the overtime premium. All the jobs together created the overtime. On the other hand, if it is a special rush job which will cause overtime premiums, then the job cost estimate for only that job should include the overhead premium, as a general rule.

What about idle time of workers and other unproductive time utilized for such things as cleanup, coffee breaks, and the like? Such costs should be run through overhead. Where they can be traced directly to a line (such as cleanup time) they should become part of the department or line overhead rate. Where this is not possible, they should become part of the general overhead and allocated as indicated in the following paragraphs.

Costs Deserving Special Consideration

We have enunciated the general principle that any and all costs directly associated with a particular job should be used in estimating the costs of that job, and no other, for price-bidding purposes. Because such extensive use is made in the industry of generalized costs standards when pricing jobs, companies are prone to miss costs for uncommon activities. Extras, created by requirements of a specific job, should be added to the estimated costs for that job. Pickling, polishing, baking to remove embrittlement, stop-off work, unusual protection of precious or fragile parts, special racks, conforming anodes, and special transportation are only a few examples of costs that should be scrutinized carefully to determine if provisions for them are included in the estimates. To the extent that any of these requirements represent services not normally performed or that are required at a higher than normal level, appropriate costs should be added. Such estimates ought to include the labor, materials, and overhead adequate to recover these extra expenses, since all of them are variable and directly traceable to the particular job.

If these costs are charged to overhead and amortized over all jobs, inequities occur. The shortcomings of this approach lead to the particular jobs being underpriced and all other jobs being overpriced. The overpriced jobs are then lost to companies correctly pricing their work. The underpriced jobs are obtained at a loss of profits for the company.

For many companies, costs to design and build or purchase special racks of conforming anodes are significant extra-cost activities that also should be estimated separately for each job. When the bidding company prices special racks separately, it may propose to deliver the racks to the customer at the completion of the job or retain them for other future use. In the case of the former, the job bid normally should list the special rack cost separately. If the racks are to be retained for future use, these costs should be spread over the estimated quantity of parts to be plated using them, by use of a departmental or finishing line overhead rate.

A flaw in the application of equipment-based costing occurs in several low-production and some high-production finishing companies when they fail to take into account possible nonproductive time that can result if there is a need to set up or deactivate a tank or a machine for a particular job. Good pricing should consider whether the job will need to begin from an empty tank, and also whether the tank setup might need to be "torn down" after the job. Companies with fewer repetitive jobs find that many jobs require, as a minimum, different anode and cathode bars, different anodes, and perhaps solutions of different chemicals at different temperatures. Such costs should be assigned directly to the particular job giving rise to them. Although the prevailing industry practice is to charge them to the machine use charge, which is prorated across all jobs, this practice overprices jobs with little or no setup or teardown requirements. When one is in doubt about which alternative to adopt, the choice that assigns the costs directly to the benefiting job is preferred — except when this practice creates a competitive hardship on potential jobs otherwise considered desirable, or in cases when segregating or recording costs in such detail is economically unfeasible.

Overhead

The greatest difficulties we observed in cost estimating for pricing decisions was in the area of overhead. A number of reasons account for this. First, there often was an inadequate or incorrect breakdown of overhead items between fixed and variable components. Second, direct costs of overhead frequently were not distinguished from indirect. For example, depreciation of machinery and equipment used on a finishing line is directly traceable to that line even though it is not directly traceable to a given job performed on that line. Depreciation on the building, on the other hand, is neither traceable directly to a finishing line nor to particular jobs. Such overhead items do differ in their traceability, and consequently they require different handling if accurate cost estimating and price setting are to result. The guiding principle, then, is that overhead should first be traced to the smallest cost center to which it is applicable. This means "directly" traceable—i.e., without any intervening allocations being necessary. Then overhead can be considered for control and price-setting purposes.

Assume, in order to cover a number of variables, that a given firm has two finishing lines, one automated and one not automated. Obviously, the automated line (designated line A) is capital intensive, while the nonautomated line (designated line B) is labor intensive.

The first principle to be followed is to trace all overhead items directly associated with each line to that line. The second principle is to break down overhead per line into its fixed and variable cost components. These can then be used to establish a predetermined fixed overhead rate for each line. If adequate accounting records are kept, most production costs (not including materials, labor, and special job-related costs as previously indicated) can be traced to each line. For example, depreciation on the equipment used on a line, repairs and maintenace of the machinery and equipment, water, steam, electrical power, factory supplies, and most other production costs can be traced to specific finishing lines.

After breaking these down into the fixed and variable costs (again, ideally, using anticipated costs for the coming year), the totals derived can be divided by estimated machine hours of operation to predetermine the fixed and variable overhead cost per machine hour for line A, and by estimated direct labor hours to derive a predetermined fixed and variable overhead rate per direct labor hours for line B. When pricing a job to be performed on line A, the estimated machine hours to be consumed can be multiplied by the fixed and variable predetermined rates per hour to get the figures to add in arriving at total estimated cost for each job. The same thing can be done for line B using direct labor hours.

But what about "general" overhead, which cannot be directly

traced to particular lines, such as depreciation on the building, heating of the building, and most selling, general and administrative costs? These, too, should first be allocated between fixed and variable costs (most will likely be fixed or mixed costs). Once this has been done, you then determine the estimated material, labor, and overhead costs for each line in total. Add total costs for all lines together. Divide this total into total variable "general" overhead to derive a predetermined variable overhead rate per dollar of cost. Divide the total into fixed "general" overhead to derive a predetermined fixed rate per dollar of cost. These "general" overhead costs can then be estimated per job by multiplying the sum of the estimated materials, labor and overhead costs (using the finishing line variable and fixed overhead rates) by these predetermined rates.

The general theory supporting this approach is that such costs ought to be estimated and allocated on the basis of relative sales value of work done on each line and for each job on the line. Since we are trying to estimate the asking selling price, however, we cannot do this. So we approach the problem from the cost side. Even though there is no necessary direct relationship between costs and selling price (as indicated in a previous chapter), nevertheless, as a general rule, the greater the effort expended, the greater will be the expenses incurred and the larger sales which will result. Usually, for example, there is some relationship between the size of selling expenses and the size of sales. The larger the work force, the greater the clerical work involved, etc. Consequently, though not provable in any mathematical sense, it can be argued that the total "general" overhead bears a relationship to the total costs per line and per job, and should be allocated on that basis for price-setting purposes.

Follow-up is critical. Ninety-five percent of questionnaire respondents indicated that they compared actual operating performance against the estimates used in pricing in order to get some idea how effective their price setting had been. To many respondents, however, this means dividing sales for a period, such as a month, by direct labor hours for that period. Assuming the resulting figure was greater than some total cost figure per direct labor hour, a profit resulted, and the manager had some evidence that his pricing had been effective. The larger the difference, of course, the better he felt.

Although such calculations are of some usefulness as rough approximations of pricing effectiveness, they are not good enough. Actual costs incurred must be accumulated per job and these compared with the estimates that were made when bidding on the job. Such information not only will aid in helping the manager improve his estimating procedures, but such data also can act as a basis for implementing cost control techniques and for budgeting and forecasting purposes.

Conclusions

We recognize that many firms will not be able to implement these suggestions overnight. Accounting systems which break down sales and expenses by line and by job are necessary. Costs must be separated into their fixed and variable components. Historical data must be accumulated on these bases to be used as a foundation for budgeting and forecasting. Firms which do not have such systems will have to establish them and obtain data for some period of time before our recommendations can be implemented. Certainly there is no better time to start than now!

We also recognize that these suggestions might have to be modified by individual firms considering their particular conditions and circumstances. Nevertheless, the general principles, we feel, are applicable to most if not all firms and should not be ignored. If an individual firm cannot adopt all the suggestions, the closer it can come, the better. In a word, the principles should act as standards for gauging performance.

VIII

CONCLUSIONS AND RECOMMENDATIONS

We hope the discussions to this point have not given the impression that the metal finishing industry is in its waning years. Not at all! Our charge was to concentrate on the problems that exist in order to help firms individually, and the industry in general, become more successful. All industries and all companies have problems which call for consideration, so the metal finishing industry is not unique in this respect. It is to the credit of the National Association of Metal Finishers and the Foundation that they should face these problems squarely, that they should ask such questions as "What are the criteria for a successful metal finishing firm?", and then look for answers. Our objective in this monograph was to help find those answers. Because of the nature of the industry, its operations, and the types of customers served, it does have special problems, or, at least, special aspects of certain general problems. These are the problems requiring satisfactory solution for successful operations, so we have placed primary emphasis on these unique features.

Captive vs. Independent Shops

By now, the reader is aware that little time has been spent to this point on the impact of captive shops on the operations of independent metal finishing firms. Two reasons account for this. In the first place, the questionnaire results and our field interviews indicate that captive shops are not nearly the problem they were when the 1967 study was made. Only twenty-seven percent of the respondents in our current survey indicated that captive shops are substantially reducing their business volume, a marked decrease since 1967 when captive shops were being created at a rapid rate and constituted a serious threat to the independents. Since the late 1960s, particularly during the last few years, our information indicates that only a small number of captive shops have been created. Most companies wishing to go the captive shop route had done so by 1967, and independent metal finishing firms have since adjusted to their existence. Second, our conclusions regarding the relative merits of independents and captive shops remain essentially the same as stated in the 1967 study; that document may be referred to rather than repeating all of the discussions here.

Even though captive shops don't pose the threat to the industry in general that they once did, in certain specific areas of the United States they still have a material impact on the success or failure of operations of specific firms in those areas. Not only did they remove business that these independent firms had at the point of their creation, but when their parent companies' operations decline to the point where the productive capacity of the captive shop is not fully utilized, they compete for work usually performed by independents. Worst of all, they frequently compete by bidding for a job at less than estimated total costs, often on an out-of-pocket cost basis. Consequently, in such cases they become a part of the cut-throat competition problem we have discussed. A number of job shop managers described specific instances where, when the captive shop's productive capacity wasn't being fully utilized by the parent, the captive shop would enter the market, underbid, and take away outside business which otherwise would have been performed by the job shop. In the light of the fact that pressure is exerted by employees on captive shops not to "farm out" work when it might idle them, it appears somewhat incongruous and inconsistent for captive shops to underbid for outside work when the production of the company owning it falls off and creates idle capacity. This is a "heads I win, tails you lose" philosophy hard to justify by any reasonable standards. In any case, a brief updating and summary of the 1967 conclusions are in order, and it seems most appropriate to present them in this concluding chapter.

Statistics on the number of employees and value added by employees in captive finishing shops usually are confidential information and not available. Even where such information is not competitively sensitive, the federal government's data gathering system aggregates information on captive operations with the statistics for the parent company or industry in which it is located. For example, statistics for captive finishing operations in the appliance industry are aggregated in the data for the appliance industry.

Nevertheless, it is known that the number of captive shops do outnumber the independents, perhaps by as much as 50 to 100 percent. Certain other facts about captive shops are also known. Many captive operations are large, some employing up to several hundred workers. Because of their relatively greater financial strength, they also are more likely to use expensive automated equipment.

Average wages and fringe benefits are considerably higher in captive plants. A study conducted by the U.S. Labor Department disclosed that in December 1973, the occupation designated "Metal Finisher" in automotive manufacturing plants was paid an average of $5.48 per hour.[1] We found during this study that the average hourly rate in independent shops serving automotive markets was about $4.25 per hour in 1976. Assuming that wages were raised in the captive shops at seven percent annually, these companies would be paying about $6.71 per hour in 1976, approximately $2.50 more than the independent shops. Although automotive wages are commonly higher than those of most other industries, we found that wages paid by job shops serving these other industries also were commensurately lower.

Costs of fringe benefits can also be considerably higher in the captive shops. Such benefits range from 35 to 60 percent of the hourly wage, compared to 20 to 40 percent in the job shops. Captive shops tend also to be more tightly constrained by work rules specified in union contracts. By contrast, relatively fewer independent shops are unionized. And, when they are organized, greater flexibility is evident (for example, salaried staff serving as part-time truck drivers or hi-lo operators) due to the closer personal relationships between the few workers and the manager. Captive shops attempt to combat higher wage costs with automated equipment; thus, a less labor-intensive operation characterizes many captive shops.

Even when automation will not overcome higher labor costs, many manufacturers argue it is still economical to operate their own captive finishing operations. Increased part handling and transportation costs, production process interuptions, and greater inventory carrying costs brought on by delaying the manufacturing flow when parts are shipped outside for finishing are only a few of the expenses cited in the justification for captive operations. Also, in emerging industrialized areas of the South and West, independent shops may be too few in number or may be located too far from the manufacturer's plant to provide a rational qualified source for outside finishing.

Once the investment for a captive shop has been sunk, it becomes difficult for a manager to idle this investment after job shops spring up nearby—even though they offer prices based on labor rates that are two or more dollars per hour lower. However, if the cost comparisons include the significant differences inherent in full-costed overhead rates in the captive and independent shops (captive shops frequently have overhead rates of between 600 and 800 percent of labor costs, while independent shops are often in the 100 to 200 percent category), the potential savings lead cost-conscious managers to send overflow finish-

1. Philip M. Doyle, "Wages at Auto Assembly Plants Top Those at Parts Factories," *Monthly Labor Review,* 99 (June 1976), p. 46.

ing to the new job shop, and finally to retire a captive line when it needs refurbishing or replacement. Where environmental protection standards require installation of pollution abatement equipment costing \$100,000–\$250,000, the decision to subcontract the finishing and close aging captive operations becomes easier.

On a full cost basis, therefore, the traditional advantages still remain with independent metal finishing firms. While captive shops basically are dependent upon one customer, job shops typically have a variety of customers from a number of different industries (although work still needs to be done in this area of market diversification, as we discussed). Captive shops are tied to the activity cycle of one type of business, operating at a low percentage of capacity when this business is operating at a low level, and vice versa. Job shops, having as customers a variety of businesses with a variety of business cycles, usually are able to operate at a higher average percentage of capacity. Because of their substantial investment in equipment and facilities, this means the independent job shop's average overhead cost per unit is likely to be lower.

Temporary advantages, such as the utilization of idle facilities, should not be allowed to cloud long-range considerations. Admittedly, once a captive shop is established it is difficult, in the shortrun, to argue that the captive shop should be closed down as long as the price charged by job shops exceeds the out-of-pocket costs of the captive shop. Although we found cases where even this was true because of the higher wage rates and fringe benefits of the captive shops, independents generally cannot operate on the basis of a selling price less than the out-of-pocket costs for captives. We found that job shops often are able to perform finishing work profitably at a price lower than total costs for a captive shop, however.

Owners of captive shops must keep in mind that decisions to continue operations when total costs of their operations exceed the prices at which the work would be performed by job shops are short-run decisions. Over the long run, plant and equipment must be replaced. Over the long run, therefore, if total costs of captive shops exceed the price at which the work could and would be performed by job shops, then the manufacturing company would decrease its long-run profits and rates of return by maintaining captive shop operations. Under such conditions, the captive shop operations should unquestionably be liquidated over a period of time.

But this comparison of total costs of captive shops to the selling price of independents is too severe a test. Although, in many cases, job shops could compare favorably even under those severe conditions, the real economic test is that the use of full costs plus an attractive rate of return on capital investment must be used to justify a capital shop facility over the longrun. Plant capacity as well as machinery and equipment will have to be replaced at some future date, making use of funds which could be utilized for other investment opportunities. Therefore, unless the criterion of full costs plus a rate of return on investment at least equal to that which could be earned on the best alternative use is used for comparison purposes, the manufacturing plants will not gain full advantage of long-range maximization.

In sum, labor and overhead costs are likely to be lower in efficiently run independent shops than in captive shops. Such reduced costs usually will create an advantage in favor of independent shops as compared to captive shops. These considerations, plus the fact that fairly stringent federal pollution abatement rules are likely to be invoked more universally within the next year or two, make it an opportune time for manufacturing firms to undertake some relative cost studies (make-or-buy type studies) of captives vs. independents. We feel in many cases their conclusion will be to phase out the captive's operation. We recognize that this cannot be done in those cases where the size of particular items to be plated or the required volume of production exceeds the capacity of independents in the area to handle the job. But in other cases, we feel manufacturing concerns will conclude that higher long-range profits will result if they have metal finishing work done outside. In any case, we feel that captive shop activity, at best, is likely to decrease in the future; at worst, we cannot foresee any substantial increased activity in this area.

Public Relations

We would be remiss if we completed this monograph without a brief discussion of the public relations problems of the metal finishing industry. As Mr. P. Peter Kovatis, executive director of the National Association of Metal Finishers, has said, "Unlike many traders, electroplating and metal finishing lack the familiar ring of a household word."[2] Worse than not being a household word, the industry probably is presently receiving more that its fair share of criticism for damaging the ecology. These things make it more difficult to attract outstanding talent into the industry and to do other things necessary for success.

Improving a public image is difficult. Attitudes develop slowly, and are just as slow to change. Public relations, therefore, must become an ongoing, continuous process. It is a job never really completed.

We have made a number of recommendations which should also help in the public relations area. Sending brochures to colleges and universities and recruiting at their campuses will aid in making the industry better known and understood. Projects such as this should

2. P. Peter Kovatis, "NAMF Official Wants to Make Metal Finishing a Household Word," *Finishers' Management* (May 1972) p. 15.

help. The results are read by many outside the industry; students frequently make use of the materials in courses they are taking, and the publisher publicizes the monograph's availability to large numbers of people within and outside the industry. The very fact that the Foundation is willing to undertake an investigation that will aid consumers and the economy over the long run by increasing effective competition is good public relations.

Mr. Kovatis, in the article referred to, suggested that the organizations representing the total industry should pool their resources for a bona fide public relations effort to make more people aware of the metal finishing industry and to attract more and better qualified workers to the field. Such efforts should help. When a firm or firms in a given area do an effective job of combating pollution, every effort should be made to publicize these efforts in local newspapers and on radio and television programs.

What Individual Firms Must Do For Success

If the whole industry is to be more successful, individual firms must, on the average, become more successful. The National Association of Metal Finishers and the Metal Finishers' Foundation can do much to help, as we will discuss presently. The attempt at problem solving, however, and the major effort must be at the individual firm level. The Foundation and National Association can be of help, but *help* becomes an exercise in futility if the managers of individual firms are unable or unwilling to make strong individual efforts to better their performance.

We have discussed the following in some detail throughout the monograph. Here we simply state them in an organized summary fashion. First, however, a word or two of caution is in order. Although we have found that the following suggestions and recommendations are materially significant to successful operations — and that many firms are weak in one or more of these conditions or requisites for success — we do not suggest that all are applicable to each and every firm in the industry. But each manager must determine as objectively as possible whether or not the suggestions are applicable to his firm and take corrective action where warranted.

In the following statements we frequently refer to the "manager." We recognize that one individual cannot perform all of these functions well, and we have suggested in preceding chapters that he avoid attempting to do so. He might obtain competent outside help or develop a management team internally that can perform all the necessary management functions efficiently. When we use the term "manager," we mean it in this sense of "management team" or use of "outside consultants and experts."

Making sound management decisions is, of course, the basis for a successful firm. Everything that needs to be done, and the way everything ought to be done, is encompassed in that phrase "making sound management decisions." Prerequisites for making the right decisions are obtaining the needed information regarding the firm, understanding it well enough to know what it means and how it can be appropriately used, and then using it to choose from alternatives the one or ones most beneficial for the firm. *Consequently, we feel the basic need for successful operations is an adequate amount of the right kind of accounting information, properly utilized to make correct business decisions.* Perhaps more basic than this is the need for many managers to become convinced that the benefits to be gained from such information exceeds the costs of obtaining it.

Mainly this can come about through education. Seminars, self-study programs, and executive development programs are useful in convincing managers that accounting information, in adequate form and amount, can aid in making better pricing decisions, in controlling costs more effectively, in utilizing assets more effectively and efficiently, in acquiring capital funds, and in improving the quality of work performed—in a word, in maximizing long-run profits. Such seminars would not only convince managers that accounting information can help increase profits but also show them how it can be done.

Following from this, we found that greater use needs to be made of cost analyses for product pricing decisions. Contribution margin analyses are needed to determine the best mix of finishing jobs, where additional marketing (sales) effort is needed, and those jobs and customers that should be dropped (or not accepted) because they do not make a sufficient contribution to profits. Budgeting techniques need to be utilized both in pricing and cost control. Managers need to utilize the funds statement much more than has been the case in the past, and to make maximum use of income statement and balance sheet data.

Evolving from analyses and understanding of accounting reports, other requisites for success become apparent. Managers need to delegate responsibility and the commensurate authority to a greater extent. Being independent and aggressive individuals on the average, metal finishing firm managers find delegating responsibility difficult to do. Nevertheless, it needs to be done. Many management functions need to be performed well (technical production, personnel management, marketing, financial analysis), and one person is not likely to be able to perform all of these at the level needed for successful operations, no matter how small the firm may be.

Technical production activities frequently are handled best of all management responsibilities, due in large part to the engineering and technical backgrounds of many managers, However, improvements still need to be made in terms of plant layout to obtain a more orderly flow of jobs throughout the plant, in terms of utilizing materials more efficiently, and in keeping the plant cleaner and better organized with respect to storage of materials and finished jobs.

Markets should be expanded by many firms to include a number of different industries and a variety of customers within each industry, rather than depending on one or two large customers in one industry.

Personnel management techniques need to be learned and applied by finishing firm managers if the productivity of workers is to reach and be maintained at a high level.

Capital improvements need to be made in many cases; new technology must be learned and adopted. Managers need to be involved in greater research and development activities.

Responsibility of National Organizations for Success

There are major reasons why it is difficult to solve industry and firm problems satisfactorily. There are reasons why success is not easily achieved. Many firms are small, and the costs of budgeting, research, and accounting analyses that are necessary for efficient and effective operations may be prohibitive for them individually. The main characteristic of the problems are not peculiar to the individual firm, i.e., they are regional or national in scope. An approach to solutions to many problems, therefore, must be made on a broad front.

The Metal Finishers' Foundation and the National Association of Metal Finishers have done excellent work, during the past fifteen or twenty years in particular, in providing useful information to member firms, in developing and conducting seminars, and in representing the interests of the members in a myriad of other ways. As always seems to be the case, however, more needs to be done.

1. A recruting effort to attract young talent into the industry, on a national scale, needs to be undertaken.
2. Thought should be given to increasing the developments of centralized research activities. Technological advance is essential to the future success of metal finishing firms.
3. Develop and conduct seminars on pricing, cost analyses, personnel management, and on using accounting data effectively — all specifically related to the problems of the metal finishing industry.
4. Develop "normal" ratios and other accounting relationships from data gathered from successful firms in the industry. These can be used as standards by individual firms to gauge performance.
5. Conduct market research and promotional activities.

6. Act as a training center for foremen, managers, and technicians where programs could be directed primarily at problems peculiar to an area.
7. Develop a plan to increase membership in the National Association. Once in the national association, finishing firm managers can be encouraged to attend seminars and otherwise participate actively in order to gain the knowledge needed to operate successfully.
8. Although the Uniform System of Accounts prepared for the National Association of Metal Finishers is useful in a number of ways, the expense accounts should be further designated as fixed or variable. Although a high degree of perfection in this regard cannot be expected, we know that successful managers can make a useful breakdown for aiding in pricing decisions, cost control, and the preparation of direct costing statements. Such a breakdown, through rough, would provide information that would encourage many firms to take the necessary first steps toward following the recommendations made in this monograph. After the initial step, managers can then worry about making refinements.

Concluding Remarks

Many of the problems lie with job shop managers themselves. Some are inept in price determination practices; some accept jobs which cannot be delivered on time; some simply are not good managers. Most such shops probably are no longer in existence; certainly they will not be in business for any extended period of time if more severe national pollution abatement requirements come into existence, (which seems likely). Improvements in overall management techniques can be made, we feel, following the suggestions and recommendations contained throughout this report. These are needed and necessary for successful operations. We are optimistic that such improvements will take place, particularly considering the competitive pressures.

We would like to emphasize that the existing problems need not portend a dismal picture for the future of the independent metal finishing firms. The finishing industry is not peculiar in having problems; all industries have them. They must be attacked with vigor, and long hours and much effort will be required before adequate solutions are found. Luck certainly plays a part in success, but the main requisities for success are observing the problems which exist, assuming that there are possible solutions, and then putting in whatever time and effort is necessary to find those solutions. Because of the strong, individualistic, competitive spirit we observed throughout the industry, we see a bright future.